AF660073

Innovations in Fruit Crop Cultivation for Enhancing Yields and Quality

NIPA® GENX ELECTRONIC RESOURCES & SOLUTIONS P. LTD.
New Delhi-110 034

About the Authors

Parmar Virat M., seniorauthor of this book received B.Sc. (Hons.) degree in Horticulture in 2018 and M.Sc. degree in Fruit Science in 2020. Currently, Pursuing Ph.D. in Fruit Science during 2023. He is actively engaged in research activities. He has published several research papers, review papers and book chapters to this credit. He has also presented her research papers in National and International seminars/conferences.

Lunagariya Radhika J., seniorauthor of this book received B.Sc. (Hons.) degree in Horticulture in 2020 and M.Sc. degree in Fruit Science in 2022. Currently, Pursuing Ph.D. in Fruit Science during 2023. She is actively engaged in research activities. She has published several research papers, review papers and book chapters to this credit. She has also presented her research papers in National and International seminars/conferences.

Dr. D. R. Kanzaria, senior author of this book received M.Sc. (Horti) and Ph.D. degree in Horticulture in 1999 and 2015, respectively. He is actively engaged in teaching of U.G. and P.G. programmes, research, extension activities and has guided many post-graduate students. He has to his credit number of research papers, popular articles and three books and received several awards. He has also presented his research papers in many National and International seminars/conferences.

Innovations in Fruit Crop Cultivation for Enhancing Yields and Quality

Parmar Virat M.
Department of Fruit Science
College of Horticulture
Junagadh Agricultural University
Junagadh – 362001, Gujarat, India

Lunagariya Radhika J.
Department of Fruit Science
College of Horticulture,
Junagadh Agricultural University
Junagadh – 362001, Gujarat, India

D. R. Kanzaria
Associate Professor
Department of Fruit Science
College of Horticulture
Junagadh Agricultural University
Junagadh - 362001, Gujarat, India

NIPA® GENX ELECTRONIC RESOURCES & SOLUTIONS P. LTD.
New Delhi-110 034

Disclaimer

Information contained in this book has been published by Lambert Publishing Academy and obtained by its authors believed to be reliable and are correct to the best of their knowledge. However, the publisher and its authors shall in no event be liable for any errors, omissions or damage arising out of use of this information and specially disclaim any implied warranties or merchantability or fitness for any particular use. The information in this book does not warrant or assume any legal liability or responsibility for the accuracy, completeness or usefulness of the courseware contents.

NIPA® GENX ELECTRONIC RESOURCES & SOLUTIONS P. LTD.

101,103, Vikas Surya Plaza, CU Block
L.S.C. Market, Pitam Pura, New Delhi-110 034
Ph : +91-11-43860225, Mob.: +91 9717133558, 9540816132
E-mail: newindiapublishingagency@gmail.com
Website: www.nipaersources.com

© 2025, Publisher

Print ISBN: 978-93-58877-64-9
e-book ISBN: 978-93-58871-71-5

All rights reserved. No part of this publication may be reproduced, stored in a retrieval system or transmitted in any form or by any means, including electronic, mechanical, photocopying recording or otherwise without the prior written permission of the publisher or the copyright holder.

This book contains information obtained from authentic and highly reliable sources. Reasonable efforts have been made to publish reliable data and information, but the author/s, editor/s and publisher cannot assume responsibility for the validity, accuracy or completeness of all materials or information published herein or the consequences of their use. The work is published with the understanding that the publisher and author/s are not attempting to render any professional services. The author/s, editor/s and publisher have attempted to trace and acknowledge the copyright holders of all material reproduced in this publication and apologize to copyright holders if permission and/or acknowledgements to publish in this form have not been taken. If any copyrighted material has not been acknowledged, please write to us and let us know so that we may rectify the error, in subsequent reprints.

Trademark Notice: NIPA®, the NIPA® logos and their presentations (the way they are written/presented) in this book are the trademarks of the publisher and hence may not be used without written permission, if copied or used without authorization, the infringer will be prosecuted as per law.

NIPA® also publishes books in a variety of electronic formats. Some content that appears in print may not be available in electronic books, and vice versa.

Composed and Designed by NIPA®.

Acknowledgements

Writing this book on modern trends in fruit production has been an immensely rewarding journey, and we wish to express our deep appreciation to everyone who has been part of this endeavor.

First and foremost, we extend our heartfelt gratitude to the farmers and orchardists who have devoted their lives to cultivating and nurturing fruit crops. Their invaluable knowledge, experience, and passion have greatly influenced the content of this book.

We are also thankful to the researchers and scientists who have dedicated their careers to advancing fruit production. Their groundbreaking research and innovative practices have significantly shaped the content of this book, and we are honored to have had access to their valuable work.

Our mentors and advisors deserve special recognition for providing guidance and encouragement throughout the writing process. Their expertise and insightful feedback have been indispensable in refining the concepts and ensuring the accuracy of the information presented.

We acknowledge the support and guidance of our colleagues and mentors, whose expertise and encouragement have played a crucial role in shaping this book. Their thoughtful suggestions and feedback have elevated its quality.

Our appreciation extends to the publishers and editorial team who believed in this project and contributed their expertise during the publishing process. Their dedication to excellence and attention to detail have been instrumental in turning the manuscript into a polished book.

To our family and friends, we express our deepest gratitude for their unwavering support and understanding throughout this endeavor. Their patience, encouragement, and belief in us have been the driving force behind our determination to bring this book to fruition.

Finally, we want to thank the readers of this book. It is our sincere hope that this work serves as a valuable resource for fruit growers, students, and enthusiasts, contributing to the advancement of fruit production worldwide.

To all those who have contributed, no matter the size of their contributions, we offer our heartfelt thanks. Without each of you, this book would not have been possible.

Thank you all for your contributions, support, and belief in the significance of "Innovations in fruit crop cultivation: enhancing yields and quality" Your collective efforts have been indispensable, and we are humbled and grateful for your unwavering support.

Authors

Credentials

The authors are thankful to various sources from where the literature has been cited and written in the references at the end of each chapter or elsewhere. Authors are rendering/ giving/ providing an extraordinary credits/ credentials to the contributors whose data, definitions, explanations, graphs, charts, drawings, photographs, abbreviations, figures, tables, formats, formulas, sentences, paragraphs or whatsoever; whichever could not written/ produced here by any mistake. Authors are really obliged to those contributors whose names, references, institutions, publications, books have been utilized here by any means or format directly or indirectly. Authors are also thankful to the government publications like books, websites, pamphlets, booklets or reports in any of the format which can be said as publications whose data has been cited hereat in any format. Authors of this book have not claimed for the any information or the explanations; this is merely a compiled document from the information available in any of the format throughout the globe. Authors have apologized for the credits which might not give to any of authors or publishers, is merely a mistake and not have any intense for the same.

Authors

Preface

Fruit production plays a crucial role in the global agricultural landscape, offering nourishing sustenance, economic prospects, and environmental advantages. Recent times have witnessed profound changes in this sector, spurred by shifting consumer preferences, technological breakthroughs, and an increasing emphasis on sustainable and productive farming methods. These contemporary developments have not only transformed the way fruits are grown, processed, marketed, and consumed but have also revolutionized the entire fruit production industry.

In recent times, the fruit production industry has experienced a significant transition towards embracing sustainability and environmentally responsible methods. This shift has been driven by mounting apprehensions about climate change and its effects on agriculture. Consequently, farmers and producers are actively adopting eco-conscious strategies, such as precision farming, which involves leveraging sensor technology, drones, and satellite imagery. By employing these advanced techniques, they aim to enhance resource efficiency, notably optimizing water and fertilizer application, minimizing wastage, and mitigating their environmental footprint.

The notion of "smart farms" has become increasingly popular, harnessing the power of the Internet of Things (IoT) and data analytics to revolutionize fruit production. Through the integration of sensors, automation, and machine learning algorithms, various facets of cultivation can be closely monitored and managed. This technological advancement empowers farmers to make well-informed decisions, leading to enhanced productivity and reduced labor expenses.

In light of evolving consumer tastes, fruit producers are expanding their product range and introducing novel varieties. Prioritizing aspects like taste, nutritional value, and convenience, breeders are developing fruits with enhanced characteristics, such as longer shelf life, resistance to diseases, and distinctive flavors. These cutting-edge fruit cultivars are designed to not only improve productivity but also elevate the overall quality of fruits, presenting fresh choices for consumers and creating promising market prospects for producers.

The book comprises seven chapters, with an emphasis on presenting tables and figures to captivate readers and provide essential information at a glance. As we delve deeper into these chapters, we will explore the latest trends in fruit production, analyzing their advantages, challenges, and potential implications for the future. By adopting these innovative techniques, fruit producers can effectively respond to changing consumer preferences, promote sustainable agricultural practices, and contribute to a healthier and more resilient fruit production sector. The book "Innovations in fruit crop cultivation: enhancing yields and quality" is expected to be highly valuable not only to researchers but also to educators, students, planners, farmers, and individuals seeking to enhance fruit production quality. Moreover, it aims to serve as a valuable resource for fruit producers worldwide, offering the most up-to-date information on fruit production technologies.

Authors

Contents

List of Contributors

D.K. Varu, College of Horticulture, Junagadh Agricultural University, Junagadh - 362001 Gujarat, India

D.R. Kanzaria, Department of Fruit Science, College of Horticulture, Junagadh Agricultural University, Junagadh - 362001, Gujarat, India

F. P. Karagatiya, College of Horticulture, Junagadh Agricultural University Junagadh - 362001 Gujarat, India.

H.J. Senjaliya, College of Horticulture, Junagadh Agricultural University, Junagadh - 362001 Gujarat, India

H.N. Patel College of Horticulture, Junagadh Agricultural University, Junagadh - 362001 Gujarat, India

J.S. Parsana College of Horticulture, Junagadh Agricultural University, Junagadh - 362001 Gujarat, India

Khyati Savani Department of Biotechnology, College of Agriculture, Junagadh Agricultural University, Junagadh - 362001, Gujarat, India

Lunagariya Radhika J. Department of Fruit Science, College of Horticulture, Junagadh Agricultural University, Junagadh - 362001, Gujarat, India.

M.S. Tondihal Department of Fruit Science, College of Horticulture, Junagadh Agricultural University, Junagadh - 362001, Gujarat, India

Makavana Payal C. Department of Horticulture, Anand Agricultural University Anand - 388110, Gujarat, India

Mithapara Kinnari D. Department of Fruit Science, College of Horticulture Junagadh Agricultural University, Junagadh - 362001, Gujarat, India.

Subhrajyoti Mishra, Department of Fruit Science, Institute of Agricultural Science Siksha 'O' Anisandhan University, Bhubaneswar, Odisha, India

V.M. Parmar Department of Fruit Science, College of Horticulture, Junagadh Agricultural University, Junagadh - 362001, Gujarat, India

Vrutti Patel, Fruit Research Station, Navsari Agricultural University, Gandevi – 396450 Gujarat, India

Y Desai, Department of Fruit Science, ASPEE College of Horticulture and Forestry Navsari Agricultural University, Navsari – 396450, Gujarat, India.

List of Symbols and Abbreviations

%	Per cent
/	Per
@	At the rate of
0	Degree
^{0}C	Degree Celsius
&	And
×	Multiply by
()	Bracket parenthesis
[ ]	Square bracket
;	Semi colon
:	Hyphen
=	Equal to
-	Minus
+	Plus
FAOSTAT	Food and Agriculture Organization Statics
e.g.	For Example
PRSV	Papaya Ring Spot Virus
Ch. No.	Chromosome Number
cm	Centimeter
PFDCs	Precision Farming Development Centers
IARI	Indian Institute of Agricultural Research
UAS	University of Agricultural Sciences
IIT	Indian Institute of Technology
NAU	Navsari Agricultural University
TNAU	Tamil Nadu Agricultural University
CISH	Central Institute for Subtropical Horticulture
et al.	*et allii*; and co-workes
etc.	Etcetera
cv.	Cultivar
LAI	Leaf Area Index
FYM	Farm Yard Manure
G	Gram

GA_3	Gibberellic acid
GIS	Geographic Information Systems
GPS	Global Positioning System
HDP	High Density Planting
i.e.	That is
IBA	Indol-3-butyric acid
IAA	Indole-3-acetic acid
NDVI	Normalised Vegetation Index
US DoD	United State Department of Defense
ICAR	Indian Counseling of Agricultural Research
INM	Integrated Nutrient Management
IoT	Internet of Things
N	Nitrogen
P	Phosphorus
K	Potassium
kg	Kilogram
KNO_3	Potassium nitrate
M	Meter
m^2	Meter Square
ESA	European Space Agency
RTK	Real Tie Kinematics
DNA	Deoxyribo nucleic acid
RNA	Ribonucleic acid
mRNA	Messenger RNA
NAA	Naphthalene acetic acid
PV	Precision Viticulture
No.	Number
NRS	Non-Reducing Sugar
C:N	Carbon:Nitrogen
PF	Precision Farming
PGR	Plant Growth Regulators
2, 4-D	2,4-Dichlorophenoxyacetic acid
RH	Relative humidity
RDI	Regulated Deficit Irrigation
ET	Evapotranspiration
PRD	Partial Root zone Drying
Spp.	Species
Sr. No.	Serial Number
SSNM	Site-Specific Nutrient Management

ABA	Abscisic Acid
TSS	Total Soluble Sugar
°F	Degree Fahrenheit
UHDP	Ultra-High Density Planting
var.	Variety
viz.,	Namely
VRT	Variable Rate Technology
dS/m	Decisiemens per metre
WUE	Water Use Efficiency
YM	Yield Monitoring
<	Less Than
>	More Than
O_2	Oxygen
CO_2	Carbon Dioxide
PSB	Phosphorus Solubilizing Bacteria
LCC	Leaf Color Chart
UAV	Unmanned Aerial Vehicle
Agbots	Agricultural Robots
CFB	Corrugated Fibre Board
HDPE	High Density Polyethylene
MA	Modified Atmosphere
CA	Controlled Atmosphere
L	Litre
Zn	Zinc
SPAD	Soil and Plant Analysis Development
NDDB	National Dairy Development Board
HOPCOMS	Horticultural Producers' Cooperative Marketing and Processing Society
PP	Polypropylene
BOPP	Biaxially Oriented Polypropylene

National Dairy Development Board

1

Scenario of Fruit Production

V.M. Parmar, Vrutti Patel and Khyati Savani

Introduction

Fruit crops have been cultivated for centuries, both commercially and in amateur gardens, and many species covering a wide range of trees, shrubs, and non-woody perennials.Fruit crop species have certain common characteristics: they are long-lived, usually highly heterozygous, and clonally propagated from an elite mother plant. Fruits are increasingly recognized as an important component of a healthy diet, and consumption of temperate fruit and fruit products, especially juices and dairy derivatives, has risen sharply over the past decade.China is the largest producer of fruit crops with approximately.

Rank	Country
1.	China
2.	India
3.	USA
4.	Italy
5.	Spain
6.	Mexico
7.	Iran
8.	Philippines

Source: FAOSTAT, 2019

- India ranks the second in fruit production.
- India has witnessed increase in horticulture production over the last few years. Significant progress has been made in area expansion resulting in higher production.
- Over the last decade, the area under horticulture grew by 2.6% per annum and annual production increased by 4.8%.
- Production of fruit crops has increased from 50.9 Million Tonnes to 97.35 Million Tonnes since 2004-05 to 2017-18.
- Total production of fruits is highest in case of Andhra Pradesh followed by Maharashtra.

1.1 State wise area and production of fruit crops (Area in '000 ha, Production in '000 MT) 2021-22 (1st Advance Estimates)

Sr. No.	States	Area	Production
1.	Andhra Pradesh	757.24	18010.39
2.	Arunachal Pradesh	48.14	125.84
3.	Assam	159.98	2504.19
4.	Bihar	363.77	4986.74
5.	Chhatisgarh	221.93	2471.09
6.	Gujarat	426.11	8243.67
7.	Haryana	82.00	1398.97
8.	Himachal Pradesh	234.78	796.79
9.	Jammu & Kashmir	335.02	2237.87
10.	Jharkhand	105.64	1293.55
11.	Karnataka	452.74	8549.03
12.	Kerala	313.91	1939.45
13.	Madhya Pradesh	348.13	6141.15
14.	Maharashtra	840.86	12296.11
15.	Manipur	41.62	463.86
16.	Meghalaya	37.05	376.02
17.	Mizoram	65.97	345.36
18.	Nagaland	34.01	316.59
19.	Odisha	366.64	2782.05
20.	Punjab	102.75	2202.10
21.	Rajsthan	69.89	938.75
22.	Sikkim	20.16	50.76
23.	Tamil Nadu	320.10	5823.88
24.	Telangana	177.81	2309.96
25.	Tripura	55.45	571.43
26.	Uttar Pradesh	505.13	11259.24
27.	Uttarkhand	181.07	648.90
28.	West Bengal	283.56	3717.98
29.	Others	15.87	121.86
30.	All India Total	6967.32	-

(*Source:* Anonymous, 2021-22)

1.2 Area and production of fruit crops in all India (Area in '000 ha, Production in '000 MT) 2021-2022 (1st Advance Estimates)

Fruits	Area	Production
Almond	10	11
Aonla	100	1206
Apple	313	2437
Banana	880	32454
Ber	53	570
Lime	322	3517
Mandarin	473	6265
Sweet orange	220	3894
Custard apple	45	387
Grapes	162	3490
Guava	307	4516
Jackfruit	187	1877
Kiwi	5	17
Litchi	98	724
Mango	2339	20336
Papaya	149	5744
Passion fruit	11	57
Peach	18	108
Pear	42	279
Pecanut	1	0
Pineapple	106	1808
Plum	23	78
Pomegranate	282	3216
Sapota	80	834
Strawberry	3	14
Walnut	109	282
Total fruits	6967	102924

(*Source*: Anonymous, 2021-22)

1.3 Highest producing countries of major fruit crops

Fruits	Production
Almond	USA
Aonla	India
Apple	China
Banana	India
Ber	China

Lime	China
Mandarin	Italy
Sweet orange	Brazil
Custard apple	India
Grapes	China
Guava	India
Jackfruit	India
Kiwi	China
Litchi	China
Mango	India
Papaya	India
Passion fruit	Brazil
Peach	China
Pear	China
Pecanut	America
Pineapple	Nigeria
Plum	China
Pomegranate	Iran
Sapota	India
Strawberry	China
Walnut	China

1.4 Production share of leading fruit producing states

Sr. No.	States	% Share
1.	Andhra Pradesh	15.63
2.	Maharashtra	12.05
3.	Uttar Pradesh	10.82
4.	Gujarat	9.24
5.	Madhya Pradesh	7.62
6.	Karnataka	7.40
7.	Tamil Nadu	5.83
8.	Bihar	5.26
9.	West Bengal	3.96
10.	Chhatisgarh	19.54

1.5 State wise highest area and production of fruit cropsin 2021-22

Fruits	**Area**	**Production**
Aonla	U.P.	M.P.
Apple	J & K	J & K
Pear	J & K	J & K
Almond	J & K	J & K
Walnut	J & K	J & K
Kiwifruit	Arunachal Pradesh	Arunachal Pradesh
Peach	Uttarakhand	Uttarakhand
Plum	Uttarakhand	Uttarakhand
Mango	Andhra Pradesh	UP
Banana	Karnataka	MH
Papaya	Gujarat	Andhra Pradesh
Mandarin	M.P.	M.P.
Sweet orange	Andhra Pradesh	Andhra Pradesh
Lime	Gujarat	Andhra Pradesh
Grapes	MH	MH
Guava	UP	UP
Sapota	Gujarat	Gujarat
Custard apple	MH	MH
Pineapple	Assam	W. B.
Pomegranate	MH	MH
Jack fruit	Kerala	Kerala
Bael	Odisha	Odisha
Ber	M.P.	M.P.
Litchi	Bihar	Bihar
Passion fruit	Nagaland	Nagaland
Strawberry	MH	Haryana

(*Source:* Anonymous, 2021-22)

References

Anonymous (2021-22). Area Production 2021-22. Horticultural cultivation of crops, area and production Statistics Division, Director of Horticulture, Agriculture, Farmers Welfare and Cooperation Department, Government of Gujarat. Available at http://doh.gujarat.gov.in/horticulture-census.htm.

Anonymous. (2021-2022). Horticulture Statistics Division, Department of Agriculture, Cooperation & Farmers Welfare, Ministry of Agriculture & Farmers Welfare, GOI. Retrieved from https://agricoop.nic.in/en/statistics/state-level.

Food and Agriculture Organization Statics (2019).https://www.fao.org/food-agriculture-statistics/en/.

2

Recent Advances in Propagation

D.R. Kanzaria, Makavana Payal C. and Khyati Savani

Introduction

- The green revolution of the 1960s and 1970s ended chronic food deficits and while cereals still command the attention of policy makers, fruit production has surged impressively, making India the second largest global producer behind China.
- Annual growth in horticulture has seen fruit production grow at faster rate and constitute the larger segment of agriculture.
- No doubt, there has been manifold increase in area, production and productivity but when we compare the productivity of various major fruit crops with other developed countries, our productivity is very low.
- This may be mainly attributed by non-availability of quality planting material and seeds, existence of old unproductive seedling orchards, small and unirrigated land holdings, use of inferior genetic stocks/varieties/hybrids, poor management and non-adoption of package of practices, high incidence of pest and diseases, heavy pre and post-harvest losses and lack of trained human resource.
- To meet increasing demand the productivity per unit area per unit time and per unit investment has to be increased from limited land resources, where production and supply of quality planting material of elite varieties is one such important factor that can bring tremendous improvement in production, productivity and higher economic returns to farmers.
- This may be mainly attributed by non-availability of quality planting material and seeds, existence of old unproductive seedling orchards, small and unirrigated land holdings, use of inferior genetic stocks/varieties/hybrids, poor management and non-adoption of package of practices, high incidence of pest and diseases, heavy pre and post-harvest losses and lack of trained human resource.
- To meet increasing demand the productivity per unit area per unit time and per unit investment has to be increased from limited land resources,

where production and supply of quality planting material of elite varieties is one such important factor that can bring tremendous improvement in production, productivity and higher economic returns to farmers.

- The propagation of plants has been one of the fundamental occupations of mankind since civilization begins.
- Most of the selected plant would have lost or could have attained undesirable forms if they were not propagated under strict environment conditions.
- Propagation can be defined as multiplication of plants by both sexual and asexual means. It involves the control of two basically different types of developmental life cycles sexual and asexual.
- The fruit plants may be propagated directly from seeds or vegetatively. Although seed propagation of fruit plants is the easiest, generally it is not the best, because it scarcely allows for maintenance of the characters of the original plant.
- In addition, plants grown directly from seeds are slow to come into production. Vegetative propagation allows for the multiplication of quick bearing plants in conformity with the original mother plants.

2.1 Sexual propagation

- The plants which multiply through seeds as a mode of perpetuation.
- In sexual method of propagation, the sex organs of flower are involved in process like pollination and fertilization leads to the formation of seeds. Seeds are typically produced from sexual reproduction within a species may have different characteristics from its parents.
- Example: Papaya, phalsa, citrus, mangosteen, rootstocks plants of many fruit crops are commercially raised throughsexual method of propagation.
- $GA_3$100 ppm and Sand + vermicompost + Cocopeat (1:1:1,v/v/v) showed better germination parameters, growth parameters and survival percentage in acid lime (Lunagariya *et al.,* 2022).

2.1.1 Synthetic seeds

- Many crops are difficult to multiply by conventional propagation methods and improve through traditional breeding programmes
- Among the innovative techniques of micro propagation, the concept of somatic embryogenesis with synthetic seed production or artificial seed technology is very promising

- Synthetic seed is referred to as encapsulated somatic embryos, which functionally mimic seeds and can develop into seedling under suitable conditions
- In general, for seed propagated crops outstanding hybrids can be multiplied through tissue culture and propagated by using synthetic seed rapidly.

What is synthetic seed?

- Syntheticseedcan be defined as artificial encapsulation of somatic embryo, shoot buds or aggregates of cell or any tissues which has the ability to form a plant.
- Also called as artificial seed or synthetic seed.

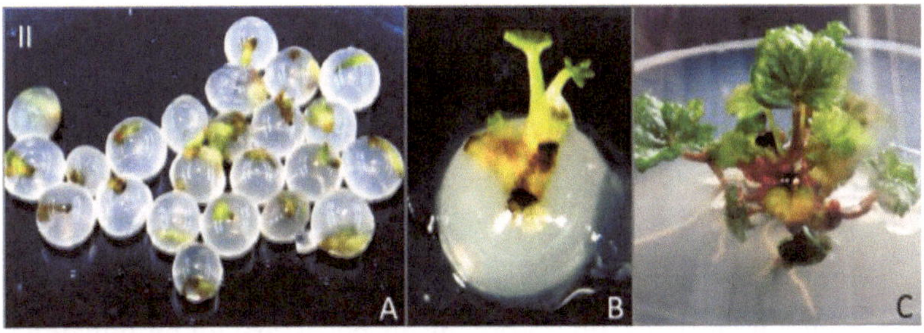

2.1.2 Somatic embryos

- Somatic embryos are bipolar structures with both apical and basal meristematic regions,which are capable of forming shoot and root respective.

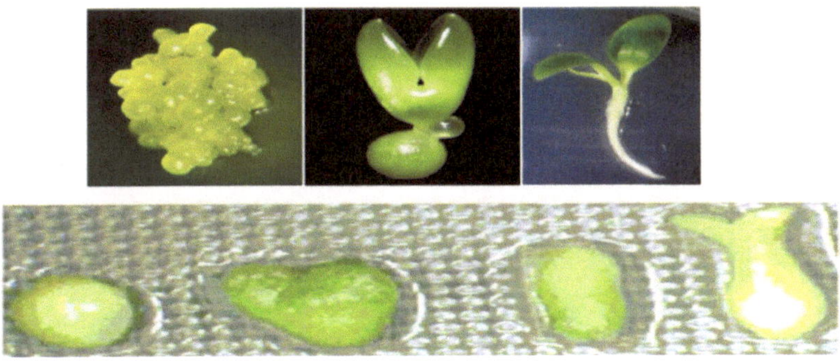

globular heart torpedo cotyledonary

2.1.3 Synthetic endosperm

- The encapsulated synthetic seed contain nutrients and growth regulators to serve as an artificial endosperm to the encapsulated propagules which results in increase efficiency of germination and viability of synthetic seed.
- MS medium most preferable.
- Along with MS media addition of adjuvants to protect and preventing the embryo from desiccation.
    - Fungicides
    - Pesticide
    - Antibiotics
    - Useful microorganism
    - Activated Charcoal
- **Desirable attributes of artificial seed coat**
    - Non damaging to the somatic embryos.
    - Durable during storage and transportation.
    - Protect the embryo while allowing for germination and conversion.
    - Contain nutrients, micro-organisms necessary for germination.
    - Sowable using existing farm machinery.

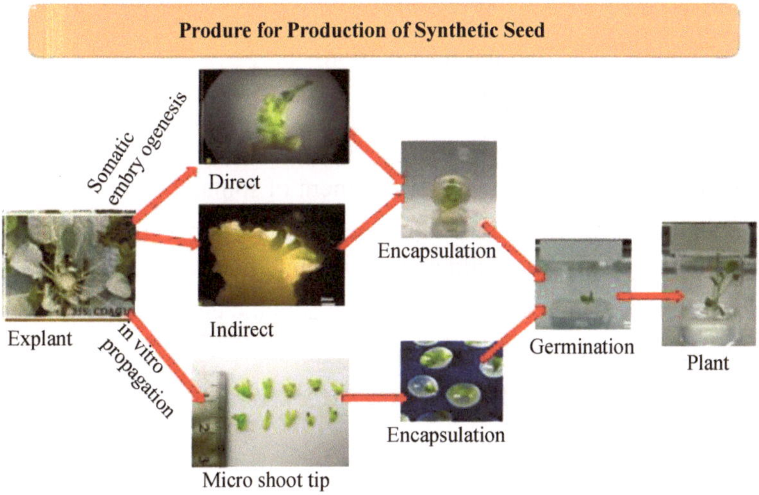

2.1.4 Types of synthetic seeds

1. **Desiccated synthetic seed**
   - Somatic embryos are first hardened to withstand desiccation and then these hardened embryos are encapsulated artificially with the use of polyoxyethylene glycol (Polyox).
   - Such types of synthetic seeds are produced only in plant species whose somatic embryos are desiccation tolerant.
   - *e.g.* carrot, celery
2. **Hydrated synthetic seed**
   - The gel used to enclose somatic embryo remains hydrated. Various water-soluble gels can be used for the purpose of encapsulation, like gel rite, guar gum, sodium alginate with gelatin. However, sodium alginate is the most suitable gel.
   - Hydrated synthetic seeds are produced in those plant species where the somatic embryos are recalcitrant and sensitive to desiccation.

- **Advantages of synthetic seed**
  - Hybrid plants can be easily propagated
  - Genetically modified plant can be propagated easily
  - Elite genotype can be preserved and propagated
  - Genetic uniformity is maintained
  - Direct delivery of tissue cultured plants to the field

  - Reduction in dormancy period
  - Pesticide can be added to the formulation
- **Limitations of synthetic seed**
  - Limited production of viable micropropagules
  - Anomalous and asynchronous development of somatic embryos
  - Improper maturation of the somatic embryos that make them inefficient for germination and conversion into normal plants.
  - Lack of dormancy and stress tolerance in somatic embryos
  - Occurrence of somaclonal variation
  - Special skills are required to carry out the work

2.2 Asexual propagation

- **Propagation by apomictic seedlings**: Apomictic seedlings are identical to their mother plants, and similar through the plants raised through other vegetative means, as it has the same genetic makeup as that of the mother plants.

  Example: Citrus, mango, apple *etc.*
- **Propagation by vegetative structure**: It is also called as vegetative propagation as it involves only vegetative parts without any sexual plant parts. The plant parts like leaf, stem, root and other root producing plant organs are used.

2.2.1 Cutting

- A cutting is a piece of vegetative tissue (stem, root or leaf) that, when placed under suitable environmental conditions, will regenerate plant parts and produce a self-sustaining plant.
- The fruit plants like Phalsa, pome-granate, lemon and grapes *etc.* are commercially propagated by cuttings.
- **Disadvantages**
  - Some plants do not root easily.
  - The desired benefit of rootstock of rootstock on a desired variety cannot be exploited.
  - Cuttings are prepared from vegetative portion of the plants such as stem, root, leaf and are classified according to plant part used.

- **Stem cutting**
  - A stem cutting is any cutting taken from the main shoot of a plant or any side shoot growing from the same plant or stem.
  - Propagation by stem cuttings is the most commonly used method to propagate many woody plants.
  - There are several types of stem cuttings.
- **Hardwood cutting**
  - Hardwood cuttings are prepared during dormant season, usually from one year old immature shoots of previous season's growth.
  - In case of deciduous plants, cuttings are made after pruning.
  - Generally, the cuttings of 15-20 cm length and having 3-5 buds are preferred depending upon species
  - *e.g.* grape, fig, pomegranate, apple *etc.*

- Semi hardwood cutting
  - Prepared from partial matured, slightly woody shoot.
  - *e.g.* Guava, Jack fruit, Lemon
- **Soft wood cutting**
  - Cuttings prepared from soft, succulent and non-lignified shoots which have not become hard or woody.
  - They are still flexible and non-woody
- **Leaf bud cutting**
  - Very few plants can be reproduced by leaf bud cutting because leaf cutting must regenerate both new root and bud tissues.
  - Not many plants have this capacity.

- **Root cutting**
  - Root cuttings can be used to propagate herbaceous plants in early winter when the plants are dormant.
  - Root cuttings are used to propagate plants that naturally produce suckers from their roots.
  - *e.g.* cherry

2.2.2 Layering

- **Airlayering**
  - In this method one-year-old, healthy and straight shoot is selected and ring of bark measuring about 2.5 to 4.0 cm just below a swollen bud is removed.
  - The cut is then surrounded sphagnum moss or any other material which can retain moisture for long period of time, can be used for this purpose and is wrapped with a polyethylene strip (200–400 gauge).
  - Both ends are tied with fine rope or rubber bands to make it practically air-tight and after 30 to 45 days, roots are formed in the aerial part of the plant.
  - The ideal time is February–March and July–August months for air-layering.
  - Litchi, Lime and sweet lime can be propagated by air layering.
  - Bio NPK consortium + Soil + FYM + Vermicompost (1:1:1, V/V/V) was found significantly superior for survivability of air layering of pomegranate (Solanki *et al.*, 2023).

- **Trenchlayering**
  - In this method, it is important to establish a permanent row of plants to be propagated.

- A branch is laid horizontally in a small trench to encourage the development of several new shoots from it.
- As these shoots grow, soil is filled around them and roots eventually develop.
- The mother plants are planted at the base of a trench at an angle of 45° in rows.
- The long and flexible stems of these plants are pegged down on the ground to form a continuous line of layered plants.
- The young shoots then arise from these plants are gradually mounded up to a depth of 15-20 cm.
- *e.g.*, Apple rootstocks (M16 and M25), cherry, plum *etc.*

Trench Layering

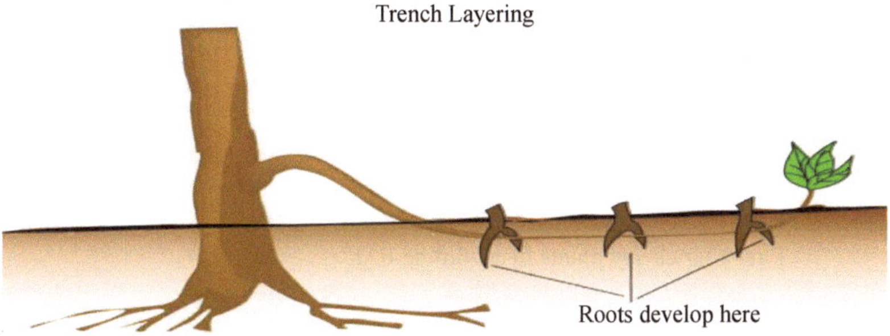

- **Tip layering**
  - In this method of propagation, bending of a plant stem to the ground and covering the tip with soil so that roots and new shoots may develop. In tip layering, the tip of shoots are bend to the ground and the rooting takes place near the tip of current season's shoot.
- **Serpentine/compound layering**
  - Serpentine layering, also known as compound layering is similar to simple layering, except that multiple sections of the stem are buried, resulting in multiple plants from one stem.
  - With this technique, a long, vine-like stem is ideal. It is modification of simple layering in which one year old branch is alternatively covered and exposed along its length.
  - *e.g.* American grape.
- **Mound layering/Stooling**
  - In this method, plants are headed back to 15-20 cm above the ground level during dormant season.

- The new shoots come out within two months after heading back.
- These sprouts are then girdled near base and rooting hormone (IBA), made in lanolin paste is applied to the upper portion of cut with moist soil.
- The rooting of shoots is observed within 25–35 days.
- The stooling is used in the commercial production of clonal rootstocks of temperate fruit crops like apple, pear etc. and sometimes in guava also.

First year cherry plants in a stooling bed

The base of the stems covered with sawdust

2.2.3 Grafting

- Grafting is another method of vegetative propagation, where two plant parts are joined together in such a manner that they unite and continue their growth as one plant *i.e.* the stock and scion are placed in close contact with each other and held together firmly, until they unite to form a composite plant.

- The different methods of grafting are tongue grafting, cleft grafting, approach grafting, side grafting and veneer grafting *etc.*
- **Tongue grafting (Splice/Whip grafting)**
    - It is the simple and popular propagation method used in apples and widely used in many fruit crops.
    - This method is commonly used when the stock and scion are of equal diameter.
    - Each scion sticks should contain at least two to three sets of buds. Identical cuts are made at the top of the rootstock and bottom of the scion, so the two pieces fit together nicely.
    - About one-year-old rootstock is headed back at a height of 23-25 cm from the soil and a diagonal cut is made at the distal end of the rootstock.
    - A similar slanting cut is made on the proximal end of the scion.
    - The cut surface of both rootstock and scion are bound together and tied firmly.
    - The scion having 2 to 3 buds is then tightly fitted with the rootstock taking care that the cambium layer of at least one side of the stock and scion unites together.
    - This is then wrapped with polyethylene strip.
    - Tongue grafting is done in March–April in high hills and dry temperate zone while February-March in lower elevation.

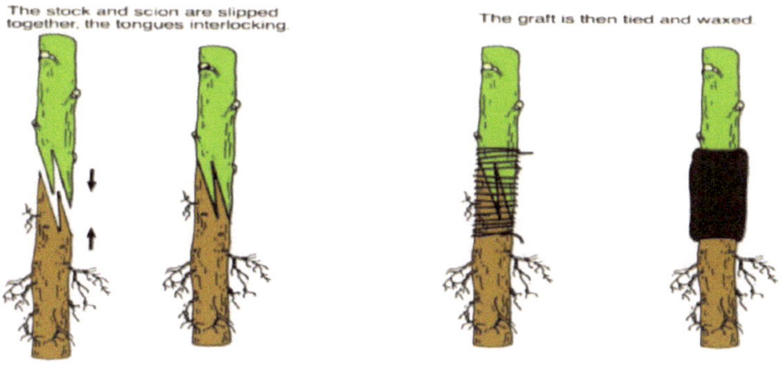

- **Cleft Grafting**
    - This is also known as wedge grafting. Wedge grafting is a relatively easy method of propagation.

    - In this technique, proper selection and preparation of scion sticks is very important for obtaining higher success of graft. This method is useful in the nursery where the rootstock is quite thicker than scion and tongue grafting cannot be employed successfully.
    - Rootstock up to 8 cm girth is selected for this purpose.
    - The rootstock is cleft grafted after decapitating the stock 45 cm above the ground level.
    - The beheaded rootstock is split to about 5cm deep through the center of stem.
    - After that a hard wooden wedge is inserted to keep open for the subsequent insertion of scion.
    - The scion of 15-20 cm size is taken from a terminal shoot, which is more than three-month-old and then it is wedge securely (6-7 cm).
    - The cleft of the scion then slipped into the split of the stock.
    - In thicker rootstock more than one scion should be inserted.
    - This technology has overcome seasonal barriers and planting materials could be raised throughout the year either in greenhouse or under open field conditions.
    - Polyethylene cap facilitates early sprouting and ensures good success rate of grafts.
    - *e.g.* avocado, apple, pear, plum, mango *etc.*
- **Side Grafting**

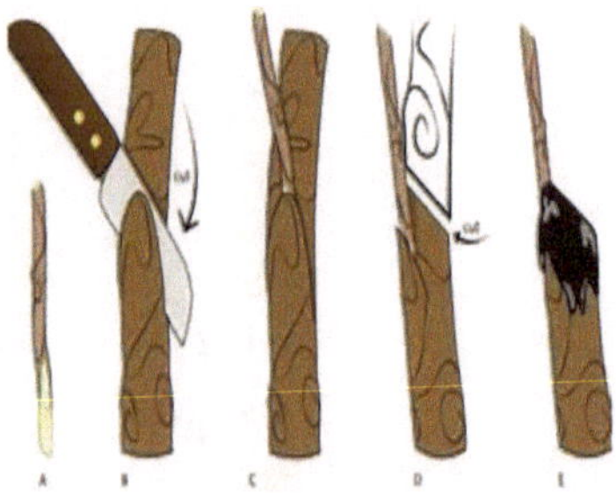

    - In this method of grafting, a three-sided rectangular cut about 4.0 x 1.25 cm is given on the rootstock at a height of about 15-18 cm and the bark of the demarcated portion is detached from the rootstock.
    - A similar cut is also given on the base of the scion stick to expose cambium.
    - The scion should be prepared well in advance before the actual grafting.
    - The healthy scion shoots of previously mature flush are selected.
    - The selected scion shoots should have plump terminal buds.
    - After the selection of the scion shoots, remove the leaf blades, leaving petioles attached to the scion.

- In about 8 to 10 days, the attached petioles drop automatically and terminal buds become swollen.
- Now the scion stick should be detached from the mother tree and grafted on the stock.
- The prepared scion is inserted under the bark flap of the rootstock after given a slanting cut so that the exposed cambium of the two components is in close contact with each other.
- The bark flap of the rootstock is resorted in its position. The graft union is then tied firmly with polythene strip.
- After the successful completion of the grafting operation, a part of the top of rootstock is removed to encourage growth of the scion.
- When the scion has shown emergence of leaves, the root stock portion above the graft union should be removed.
- Side grafting can be carried out successfully from March to October but success during the May and October is rather low. This method of propagation is commonly used in mango.

- **Inarching/Approach grafting**
  - The method of inarching or approach grafting is quite cumbersome and time consuming, but it is still the leading method for commercial propagation of many fruit plants.
  - It is generally used for repairing or replacing damaged root system and hence also called as repair grafting.

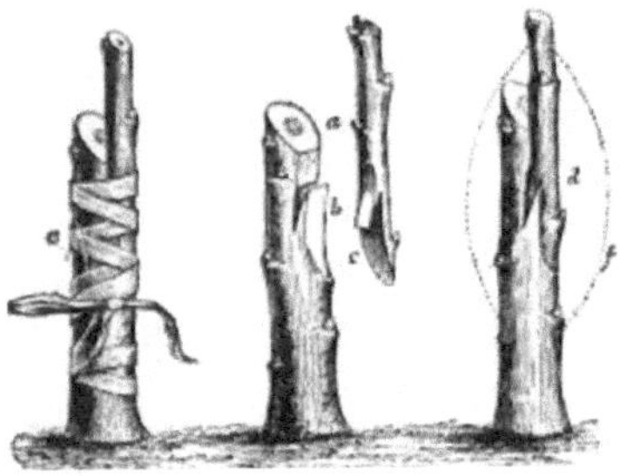

  - This method of grafting is termed approach grafting, as the rootstock is approached to the scion, while it is still attached to the mother plant.
  - Selection of parent tree for taking the scion is an important factor for its success.
  - The scion plant should be healthy, vigorous and high yielding.
  - The stock is brought close to the scion. In this method the diameter of rootstock and scion should be approximately the same.
  - A slice of bark along with a thin piece of wood about 6-8 cm long and about

  1/3 inch in thickness at height is removed from matching portions of both the stock and the scion.
  - They are then brought together making sure that their cambium layers make contact with each other.
  - These grafts are then tied firmly with polythene strip or any other tying material.
  - The stock and scion plants are watered regularly to hasten the union.
  - The union is complete in about 2 to 3 months.
  - After successful union, stock above and scion below the graft union are looped of gradually.
  - Last week of July or the first week of August is the best period for approach grafting.
  - *e.g.* mango, sapota, guava, litchi *etc.*

- **Veneer grafting**
  - This method of propagation possesses promise for mass scale commercial propagation.
  - It is simple method of propagation. However, several factors affecting the success like age and diameter of scion, season of grafting and defoliation period of scion stick *etc.*
  - About one year old rootstock is suitable for this method. However, if the stock attains suitable thickness (about 1to 1.5 cm diameter) earlier than a year, can be utilized for rootstock purpose.
  - Better success is obtained with a scion stick of 3-6 months of age with lust green leaves are selected.
  - The scion sticks are pre-defoliated 5-10 days prior the grafting leaving the petiole attached for making the auxiliary and apical buds active.
  - Usually, the terminal and next to terminal shoots are most ideal.
  - For conducting this grafting operation, a downward and inward 25–35 mm long cut is made in the stock at a height of about 15-20 cm.
  - At the base of cut, a small shorter cut is given to intersect the first so as to remove the piece of wood and bark.
  - The scion stick is given a long slanting cut on one side and a small shortcut on another side to match the cuts with the rootstock.
  - Now the scion is inserted in the stock so that the cambium layers come in contact from the longer side.

  - The graft union is then tied with polyethylene strip. After the scion remains green for more than 10 days, the rootstock should be clipped in stages.
  - When scion growth begins, the shoot of rootstock is removed above the graft union.
  - *e.g.* Mango
- **Softwood grafting**
  - In this case, as the name indicates grafting is done on newly emerged rootstock of 40-60 day-old.
  - This is easy and simple method and can be practiced throughout the year raised with 60-90% success.
  - This method is useful for in situ grafting while establishing new orchards with already established rootstocks in the field.
  - Give a transverse cut on the top soft portion of the seedling.
  - Make a cleft of 3-4 cm deep in the middle of the decapitated stem of the seedling by giving a longitudinal cut. A little wood is removed from the inner side of the cleft at the top.

  - The scion is prepares by mend the cut end of the scion into wedge shape of 3-4cm in length by chopping off the bark and little wood from the two opposite sides, taking care to retain some bark on remaining two sides.
  - Insert the wedge of the scion into the cleft of the seedling taking care that cambium layers of stock and scion come in perfect contact with each other.
  - Secure the joint firmly with polythene strip (30 cm long, 2 cm broad and 150 gauge).
  - Cover the scion with polythene cap (15 cm x 10 cm, 100 gauge) and tie it at the bottom to maintain humidity and to protect the apical bud from drying. The cap should not touch the terminal bud.
  - July and August months with high humidity and moderate temperature are the best for the success of softwood grafting.
  - This is very famous and successful propagation method in cashew nut, mango, sapota *etc.*

2.2.4 Budding

- Budding is a method of propagation in which only one bud of desired scion is inserted in the rootstock.
- The ideal condition of budding is when the bark starts slipping both on the stock and scion.
- It indicates that the cambium is active.
- Important method of budding is shield or t-budding, patch budding, chip budding, ring budding *etc.*
- **Shield/T-budding**
  - 'T' budding or shield budding is a special grafting technique in which the scion piece is reduced to a single bud.
  - A small branch with several buds suitable for T budding on it is often called a bud stick.

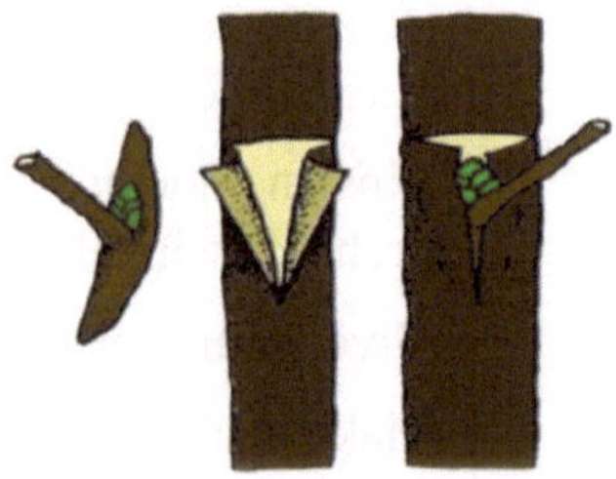

  - Leaf blades are clipped from the bud sticks, leaving the petiole intact.
  - The cut should begin about 1/2 to 3/4 inch below the bud, and should go deep enough into the wood so that when the cut is finished about 1/2 to 3/4 above the bud, the bark and a small sliver of wood are cut off.
  - A perpendicular cut across the top of the upward cut will separate it from the bud stick.
  - The bark is carefully slipped from the stem of the rootstock exposing a "pocket" into which the bud shield can be placed.
  - Care should be taken not to tear the flaps of bark in the process of spreading them. If the bark does not slip easily, this indicates that the stock is not in active growth and the process should be conducted later when active growth has resumed.
  - The bark flaps are held tightly against the bud as they are wrapped with a budding rubber, grafting tape or other suitable closure.
  - This closure must be removed in 2 to 3 weeks after the union has healed.
  - *e.g.* Citrus, plum, peach, cherry, ber, rose *etc.*

- **Patch budding**

  - A rectangular patch of bark is removed completely from the stock and replace with a similar patch of bark containing a bud of desired variety.
  - Normally done during the growing season when the bark separates readily from the wood along the cambial layer.
  - It is successfully used in species having thick bark such as walnut, pecan nut.
- **Chip budding**

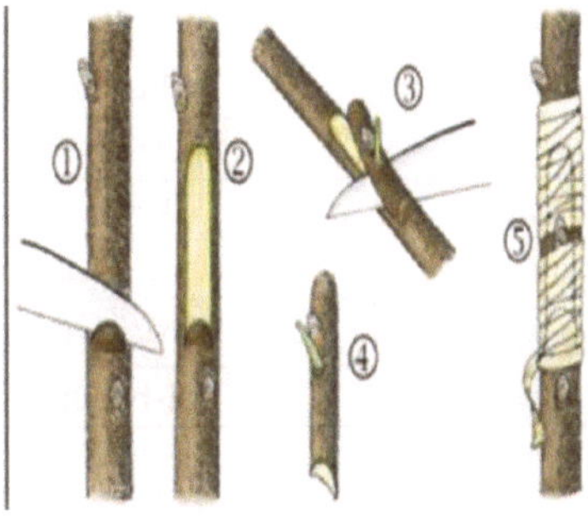

  - Chip budding is successful method of budding when the bark of the stock does not slip easily.
  - A chip of bark and wood is removed from the smooth surface between the nodes of the stock.
  - A chip of similar shape and size is then removed from the bud wood of desired cultivar.
  - For which, a 2-3 cm long down ward cut is made through the bark and slightly in to the wood of the stock.
  - The bud chip then slipped in the place of rootstock from where chip has been removed.
  - *e.g.* mango, grape *etc.*
- **Ring budding**

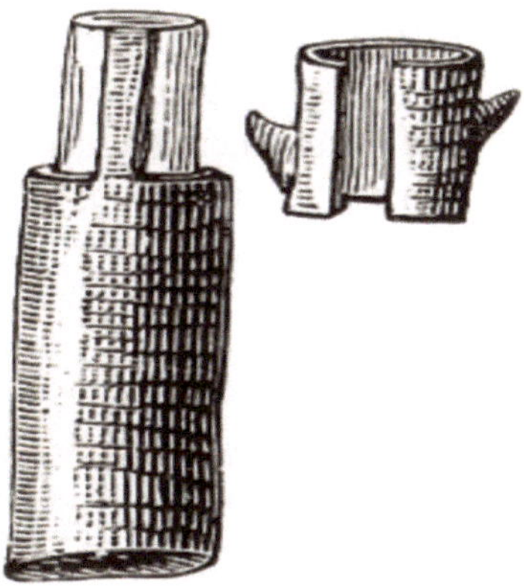

  - In ring budding, a complete ring of bark is removed from the stock and it is completely girdled.
  - A similar ring of bark containing a bud is removed from the bud stick and is inserted on to the rootstock.
  - In this budding both scion and stock should be of same size.
  - It is utilized in peach, plum, ber, mulberry *etc.*

2.2.5 Recommended propagation technique for fruit crops

Crop	Propagation method
Almond	T-budding, wedge grafting
Aonla	Patch budding, wedge grafting
Apple	T-budding, tongue grafting, wedge grafting
Apricot	T-budding, wedge grafting
Avocado	T-budding, wedge grafting
Bael	Wedge grafting
Banana	Suckers/corm, tissue culture
Cashew	Soft wood grafting
Cherry	Tongue, wedge grafting
Custard apple	Seed, Wedge grafting
Date palm	Sucker/ off shoot
Fig	Hard wood cutting/ semi-hard wood cutting
Crop	Propagation method
Grape	Hard wood cutting, wedge grafting
Guava	Air layering, wedge grafting
Jack fruit	Patch budding, soft wood grafting
Litchi	Air layering
Mandarin	T-budding
Mango	Soft wood grafting, wedge grafting,
Peach	T-budding, wedge grafting, tongue grafting
Pear	T-budding, wedge grafting, tongue grafting
Pineapple	Slip, sucker
Pomegranate	Air layering, hard wood cutting, tissue culture
Walnut	Patch budding
Sapota	Soft wood grafting, wedge grafting

2.2.6 Micro-propagation

- Micro-propagation refers to the in vitro multiplication and/or regeneration of plant material under aseptic and controlled environmental conditions to produce thousands or millions of plants for transfer to the field.
- Micro propagation is a complicated process and mainly involves 3 stages (I, II and III).
- **Stage 0:**
  - This is the initial step in micro propagation and involves the selection and growth of stock plants for about 3 months under controlled conditions.

- **Stage I**
    - In this stage, the initiation and establishment of a culture in a suitable medium are achieved. Selection of appropriate explants is important. The most commonly used explants are organs, shoot tips and axillary buds. The chosen explant is surface sterilized and washed before use.
- **Stage II**
    - It is in this stage, the major activity of micropropagation occurs in a defined culture medium. Stage II mainly involves multiplication of shoots or rapid embryo formation from the explant. A growth chamber set at 20–24 °C is used, with a 2000- to 4000-lux light intensity, and a lighting period of 16 hours or so.
- **Stage III**
    - This stage involves the transfer of shoots to a medium for rapid development into shoots. Sometimes, the shoots are directly planted in soil to develop roots. In vitro rooting of shoots is preferred while simultaneously handling a large number of species.
- **Stage IV**
    - This stage involves the establishment of plantlets in soil. This is done by transferring the plantlets of stage III from the laboratory to the environment of the greenhouse.

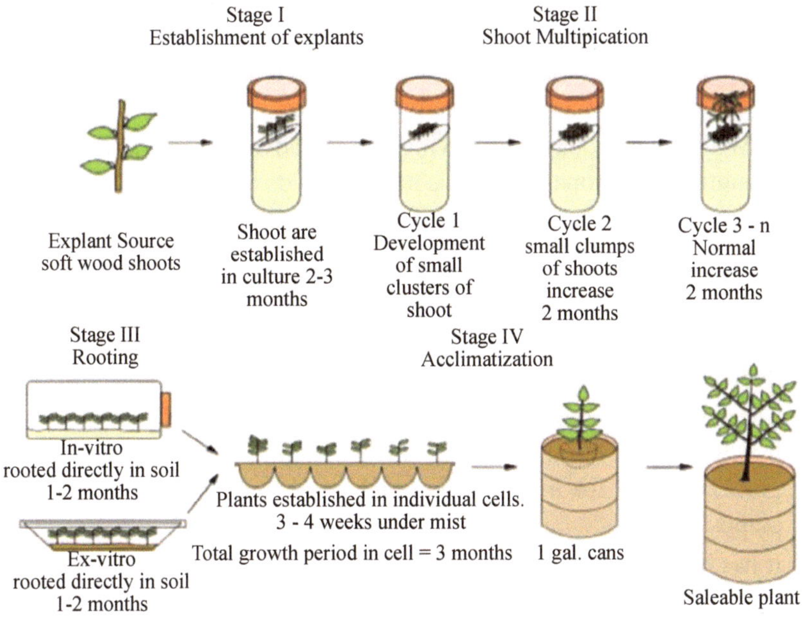

- **Application of micro-propagation**
    - Plant tissue in small amounts is sufficient for the production of millions of clones in a year using micropropagation. It would take a great deal of time to produce an equal number of plants using conventional methods.
    - The technique of micropropagation provides a good alternative for those plant species that show resistance to practices of conventional bulk propagation.
    - An alternative method of vegetative propagation for mass propagation is offered through micropropagation. Plants in large numbers can be produced in a short period. Any particular variety may be produced in large quantities and the time to develop new varieties is reduced by 50%.
    - Large amounts of plants can be maintained in small spaces. This helps to save endangered species and the storage of germplasm.
    - The micropropagation method produces plants free of diseases. Hence, disease-free varieties are obtained through this technique by using meristem tip culture.
    - Fast international exchange of plant material without the risk of disease introduction is provided. The time required for quarantine is lessened by this method.
    - The micropropagation technique is also useful for seed production in certain crops as the requirement of genetic conservation to a high degree is important for seed production.
    - With micropropagation having various advantages over conventional methods of propagation, this method holds better scope and future for the production of important plant-based phytopharmaceuticals.

2.2.7 Micrografting

- Micrografting is an in vitro grafting technique which involves the placement if meristem or shoots tip explant onto a decapitated rootstock that has been grown aseptically from seed or repropagated cultures.
- The technique was modified and improved for increasing the graft success by Murashige *et al.* (1972) and Navarro *et al.* (1975).The technique has great potential for improvement and large-scale multiplication of fruit plants.
- It has been used for the productionof virus and viroid free plants in fruit crops.

- Micrografting has also been used in prediction of incompatibility between grafting partners, histological studies, virus indexing, production of disease-free plants, safe exchange of germplasm between countries and multiplication of difficult to root plants.
- **Stages in micrografting**
  - Establishment and multiplication of scion shoot or meristem tips intended for grafting can be taken from actively growing shoots in greenhouse, chambers, field or in vitro.
  - Generally, apical shoot tips or nodal cuttings are used as explants for the establishment of *in vitro* cultures.
  - Following establishment, microshoots are transferred to shoot proliferation medium where shoot number increases by the development of new axillary shoots.
  - Microshoots of desired thickness, age and length are used as scions for *in vitro* grafting operations.
- **Establishment and multiplication of rootstock**
  - Rootstocks used for micrografting are *in vitro or in vivo* germinated seedlings and rooted or unrootedmicro propagated shoots.
  - When seedling rootstocks are used and all stages of grafting are conducted *in vitro,* seeds are surface sterilized and germinated aseptically in vessels containing nutrient salts.
  - The seedlings may be supported on agar medium. Seedlings can also be on a porous substrate, such as sterile vermiculite, which allows the growth of a branched root system.
- **Preparation of rootstock and scion for micrografting**
  - Micrografting is affected by cutting off the top of the seedling rootstocks usually just above the cotyledons or top of the micro-propagated shoot and placing small shoot apices of scion onto the exposed surface of decapitated rootstock in such a way that the cambium layer or vascular ring of the cut surfaces coincides with each other. This is called surface placement method.
  - Wedge or cleft grafting is performed, incase thickness of rootstock and scion material is large enough to allow making of wedge on the scion material.
  - Firm contact between rootstock and scion is extremely important at the graft junction for proper union of partners and callus formation (Canan *et al*., 2006).

- Several techniques have been developed for holding grafts together until fusion takes place such as translucent silicon tubing (Gebhardt and Goldbach, 1988), elastic strip (Jonard*et al.*, 1983), filter paper bridge (Huang and Millikan, 1980) and glass tubing, nylon bands, aluminum foil tubes, dual layer apparatus of aluminum foil and absorbent paper (Obeidy and Smith, 1991).
- When grafts are successful, rootstock and scion grow together to produce a plant. It is usually necessary to examine freshly grafted seedlings on a regular basis and remove any adventitious shoot arising on or below the graft union.

- **Improvement in micrografting**
    - Micrografting procedures are difficult and generally results in a low rate of successful grafts, which makes it an expensive and time-consuming production technique.
    - It is due to the fact that more technical expertise is required in preparing successful grafts on small-scale material and handling difficulties associated with preserving the delicate graft unions.
    - In many experiments, failure rate for micrografts was higher than desired.
    - *In vitro* grafts of fruit plants often fail due to incompatibility reaction, poor contact between stock and scion and phenolic browning of cut surfaces.
    - In order to alleviate some of these limitations, different techniques have been developed to make micrografting a successful and superior technology for the benefit of technicians, researchers, nursery operators and commercial tissue culture laboratories.

1. Browning and tissue blackening

- Exudation of phenolic compounds from the cut surfaces and their oxidation by polyphenoloxidase and peroxidase enzymes cause discoloration of the tissues which results in poor micrografting (Martinez *et al.*, 1979).
- Browning of the cut surfaces inhibits the growth and development of new cells and results in poor graft union.
- To block the oxidation phenomena and prevent tissue browning, various substances have been used which include thiourea, cysteine, chlorhydrate (Jonard, 1986), citric and ascorbic acid (He *et al.*, 2005).
- Tissue blackening, which commonly results in the death of very small scions, has been reduced by soaking explants in an anti-oxidant solution,

and/or placing a drop of solution onto the severed rootstock immediately before inserting the scion (Jonard*et al.*, 1990; Ramanayake and Kovoor, 1999).

2. Sucrose concentration of the medium

- Sucrose concentration of nutrient medium had a significant effect on the percentage of successful grafts.
- Navarro *et al.* (1975) reported that sucrose concentration of medium of grafted plants played a significant role and that the highest rate of successful grafts in citrus species was obtained with 7.5% sucrose.
- Generally *in vitro* growth and development increases with increased sugar concentration (Pierik, 1987).

3. Preventing desiccation of the graft

- Desiccation of graft or surfaces of the grafting partners is one of the major causes of graft union failure (Pliego and Murashige, 1987; George *et al.*, 2008).
- To prevent this phenomenon, Pliego and Murashige (1987) applied a layer of moist nutrient agar gel to connect the grafting partners and obtained better graft success.
- Micrografts in which an agar drop was added to their grafted area were highly successful (70% in apple and 60% in pear) as compared to those without an agar drop (10%).
- Adding an agar drop usually prevents scion drying and makes the transport of different materials possible and holds the graft units together until the fusion takes place.
- Addition of agar drop supplemented with minerals and/or phytohormones further improved graft success.

4. Pretreatment of shoot apex

- A technique which pre-treats the apex allowing the selection of the viable apex and helping their development greatly improves the micrografting success.
- **Application of Micro grafting**
  - Virus and viroid elimination
  - Assessment of graft incompatibility
  - Improvement of plant regeneration
  - Mass multiplication
  - Indexing viral diseases

- **Mango**
  - Amin (1974) was the first Indian who coined the term "softwood grafting" in vegetative propagation of mango. He tried this technique and succeeded in establishing *in situ* mango orchards in the arid zone of Gujarat state.
  - Grafting procedure followed for softwood grafting in mango consists of grafting on in situ rootstocks or on rootstocks raised in polythene bags for a year or more and grafting on them with scions selected from developing terminal shoots. The scion should be of same thickness as that of terminal portion of the decapitated stock. Excellent results were obtained when terminal growth of rootstock was bronze in color (Amin, 1978a).
- **Banana**
  - Commercial bananas are seedless and propagated by vegetative means.
  - The banana has a reduced underground stem, called the rhizome, which bears several buds.
  - Each of these buds sprouts and forms its own pseudostem and a new bulbous rhizome.
  - These daughter plants are called suckers.
  - Sword suckers have a well-developed base with narrow sword shaped leaves at early stages.
  - Water suckers possess broad leaves, which do not produce healthy banana clumps.
  - It can also be propagated by whole or bits of rhizomes.
  - Basraivariety inJalgaon (Maharashtra) is as a rule propagated by dormant rhizome only.
  - After cutting the parent plant, the rhizomes are removed from the soil, stored in dry place for about 2 months.
  - During the resting period, the remaining parts of pseudostem at the bottom falls off, leaving prominent heart bud.
  - The weight of rhizome should be 500-750 g. It should be 3-4 months age of planting.
  - It is commercially propagated by tissue culture. Normally disease-free plantlets with 3 to 4 leaves are generally supplied in pots for raising secondary nursery. Plants are initially kept in shade and as they harden, shade is reduced gradually.

- **Papaya**
  - Papaya is commercially propagated by seed and tissue culture plants.
  - As papaya is more commonly propagated by seeds, its cultivation is hindered by problems due to inherent heterozygosity, dioecious nature and susceptibility to several viral diseases.
  - In dioecious cultivar s of papaya, equal probability of male and female plant poses the problem of roughing excess male plants.
  - Generally, 3 seeds per pit are sown in dioecious varieties of papaya.
  - Recently, inarching and cleft grafting is also found useful to overcome the problem of sex identification.
  - Grafting and inarching of promising papaya hybrids and inbreds onto *V. cauliflora*, a wild resistant to PRSV was found to delay the symptom expression in papaya (Villegas, 1997).
  - The grafting success (about 80%) through cleft method in 'Eksotika' papaya at nursery stage stressed the advantage of grafted papaya trees as they bear fruits much lower and earlier and are dwarf in stature with longer economic life cycle (Chong *et al.*, 2007).
- **Sapota**
  - The commercial method of propagation of sapota at present is through approach grafting.
  - The root stock used for grafting is rayan belonging to same family.
  - Though it's hugely successful method but it is laborious.
  - However, there is an alternative method of propagation called soft wood grafting during the month of June-July which is easier and successful.

2.2.8 Stenting

- In Dutch language, the word 'stenting' means 'to stem' it is combination of two words '*stekken*' (to strike a cutting) and '*enter*' (to graft). In this method cutting and grafting are done simultaneously. So, the formation of the graft union and adventitious roots takes place at a same time. In fruit trees, grafting has been practised with rooted rootstock for a long time; however, use of non-rooted rootstock is relatively new.
- The traditional method of grafting takes almost two years to produce plants of saleable size. During first year the rootstock are raised through seeds or by cuttings and when these rootstocks attain the size of pencil

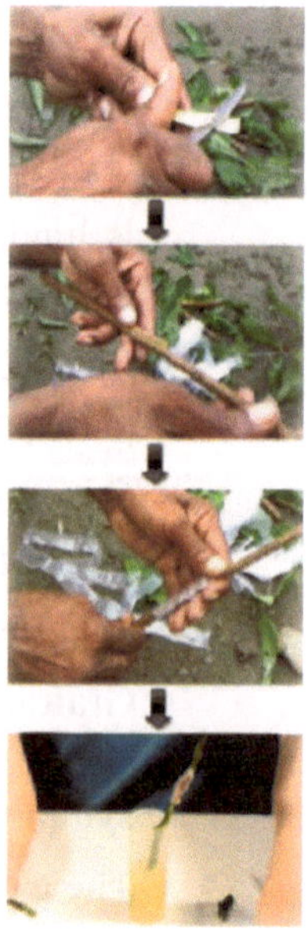

thickness (takes 10-12 months), the grafting operations are done and thus it takes another one year to produce plants for planting in the field. On the other hand, in stenting technique, graft union formation and rooting occurs simultaneously. Thus, stenting is an innovative technique to reduce the duration of multiplication, where rooting of clonal rootstock is not a complex phenomenon. Scion is grafted onto non-rooted rootstock cuttings.

- It is being used worldwide by rose growers and also a valuable technique in propagating species of conifers and also apple, rhododendron, plum and pear.
- **Advantages**
  - Stenting reduces the nursery duration. We can get a grafted plant within a year to meet the increasing demand.
  - Faster evaluation of new scion-stock combination and investigation of the interaction between shoots and roots.
  - Stenting also provides the benefits of rootstock used. The rootstock shows the positive effect on a different aspect of scion ranging from tree size, productivity, uptake of nutrient, growth rate, time of defoliation, bloom date and tree survival.
  - It is also known to affect the susceptibility of scion tree to a variety of disease and pest.
  - It helps in achieving resistance to soil salinity, nutrient imbalance, disease and pest resistance and tolerance to other abiotic condition in lesser time span.
  - Reduces the cost of nursery raising by using green house space only for a year instead of two years.
- **Disadvantages**
  - As in case of stenting sometimes success percentage is low because the graft union formation and rooting is take place at a same time.
  - This is an extensive method of propagation.
  - It requires specialized skill.
  - Life span of grafted plant is short as compared to seed propagated plant.
  - Spread of viral diseases may occur through this method.

2.2.9 Microbudding

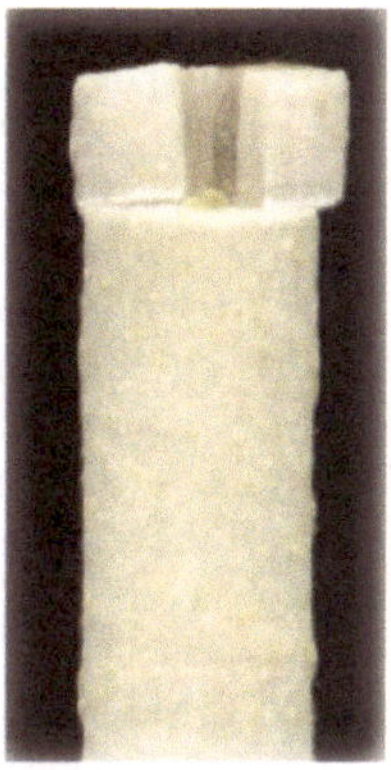

- It involves grafting a microscopic-size tip, consisting of the leaf primordia and meristem from the very end of a growing shoot of the parent stock, onto a specially conditioned toothpick-size seedling.
- Performed under aseptic conditions.
- Utilizes growing cells from the shoot tip while they are still undifferentiated and before they become a part of the underlying conductive phloem tissue that contains the virus particles.
- Two to four weeks after the initial graft, the newly emerged shoot, now established in a test tube, is ready to be regrafted onto a vigorous seedling rootstock in the greenhouse.
- A year later, after being thoroughly tested for diseases, budwood from the new plant will be available to growers to establish certified disease-free trees.

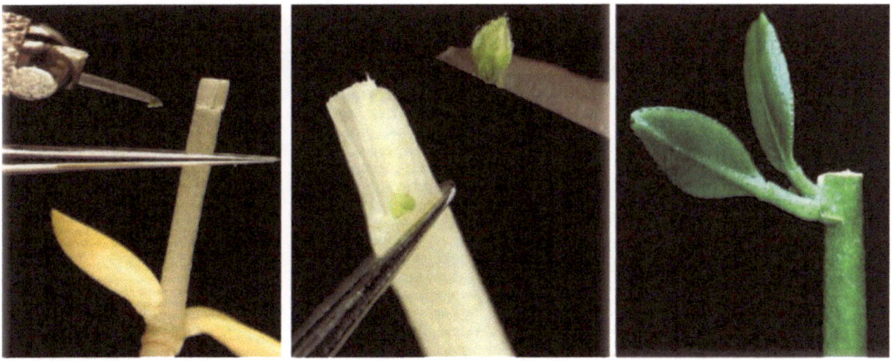

References

Canan, C.; Mehmet, O.; Hakan, T.; Kamil, S. and Elman, I. (2006). In vitro micrografting of pistachio (Pistacia vera L.) var. Siirt, on wild pistachio rootstocks. Journal of Cell and Mol. Biol.,5:25-31.

Chong, S. T. (2007). Micrografting in citrus. Acta Hort., 273: 345-349.

Gebhardt, K. and Goldbach, R. (1988). Establishment, graft union characteristics and growth of Prunus micrografts. Physiol. Plant.,72:153-115.

Huang, S. and Millikan, D. F. (1980). In vitro micrografting of apple shoot tips. Hort. Sci. 15:741-743.

Lunagariya, R. J.; Jat, R. K. and Solanki, S. M. (2022). Effect of seed soaking and growing media on germination and survivability of acid lime (Citrus aurantifolia Swingle). International Journal of Plant and Soil Science,34(23): 1720-1728.

Martinez, J.; Poessel, J. L.; Hugard, J. and Jonard, R. (1979). L'utilisation du microgreffage in vitro pour l'étude des greffes. Incompatibles.Comptes Rendus Académie des Sciences, Paris, 292(3):961-964

Murashige, T.; Bitters, W. P.; Rangan,E. M.; Naue, E. M.; Roistacher, C. N. and Holliday, P. B. (1972). A technique of shoot apex grafting and its utilization towards recovering virus-free citrus clones. HortSci.,7:118-119.

Navarro, L.; Roistacher, C. N. and Murashige, T. (1975). Improvement of shoot-tip grafting in vitro for virus-free citrus. J. Amer. Soc. Hort. Sci. 100:471-479.

Obeidy, A. A. and Smith, M. A. L. (1991). A Versatile New Tactic for Fruit Tree Micrografting. HortTechnology, 4: 91-95.

Ramanayake, S. M. S. D. and Kovoor, T. (1999). In vitro micrografting of cashew (Anacardium occidentale L.). J. Hort. Sci. Biotech.,74:265-68.

Solanki, S. M.; Jat, R. K.; Lunagariya, R. J.; Jat, M. and Kumar, M. (2023). Effect of biofertilizers and growing media on survivability of air layers of pomegranate (Punica granatum L.). Biological Forum- An International Journal,15(2): 242-245.

Villegas, V. N. 1997. In Annual Scientific Conference, Philippines, Baguio City, 1997.

3

Rootstocks Study in Fruit Crops

Lunagariya Radhika J., Mithapara Kinnari D. and Subhrajyoti Mishra

Introduction

- At present, horticulture is recognized as a potential sector to enhance agricultural production, improve house hold nutritional security and income generation through diversification and employment, value addition and export.
- Inspite of the enormous success achieved in horticulture sector, several constraints still exist.
- Besides new emerging challenges, poor productivity per unit area continues to be a concern in most of the horticultural crops with climate change impacting the productivity further.

3.1 Role of root stock in fruit crops

- The role of rootstocks and its use in different fruit crops has significant impact on fruit crop production by influencing canopy architecture, nutritional uptake, flowering, yield and fruit quality.
- Besides, it can also confront biotic and abiotic stresses such as soil pathogens, thermal stress, salinity and nutritional stress.
- Due to limited availability of arable land and high market demand for fruit crops, they are frequently cultivated under unfavorable soil and environmental conditions like thermal stress, drought, flooding, salinity and contamination of organic pollutants.
- One way to substantiate or reduce these losses in production would be the use of appropriate rootstocks, which are capable of reducing the effect of external stresses on the scion.
- Rootstocks have a primary role in determining orchard efficiency.
- They are responsible for water and mineral uptake and provide anchorage for the tree.

- Rootstocks determine tree size.
- Rootstocks can be used as interstems, a small piece inserted between the rootstock and scion in propagation process. These trees are known as "three piece" trees.
- Inter-stems can use to overcome incompatibility barriers between stock and scion and also it reduces the high vigour of desirable rootstocks.
- Rootstocks may provide some degree of tolerance to soils that are sometimes too wet or too dry.

3.1.1 Role of rootstocks in mango

- Vigour management plays an important role in mango, especially for high density planting and orchard management in terms of canopy management, harvesting and plant protection measures.
- According to Jauhari *et al.* (1977) growth and bearing habit of 'Dashehari' mango on the seedling stock of 'Dashehari' itself, showed most vigorous nature and highest yield compare to other polyembryonic rootstocks.
- Similar results were reported from IIHR- Bangalore, based on 21 years study on the performance of 'Alphonso' which indicated nucellar seedlings of 'Muvandan', 'Bappakai' and 'Olour' were vigorous rootstocks in decreasing order of vigour while 'Vellaikulumban' seedling imparted dwarfing in comparison to 'Alphonso' grafted on its own seedling (Reddy *et al.*, 2003).
- Smith *et al.*, 2008 reported that yield and yield efficiency of 'Kensington Pride' was best on 'Sg. Siput' rootstock and it was poor with Sabre rootstock indicated the possibilities for manipulating mango scion productivity through rootstock genotypes.
- Quijada *et al.*, 2008 found that 'Sinamaica' rootstock have the best adaptability to agro-ecological conditions of Maracaibo plain of Venezuela and produce higher yield and production efficiency with respect to different cultivars (Criollode Mara, Manzana and Sensation mango).
- 'Bappakai' was found to be best rootstock for 'Dashehari' followed by 'Muvandan' and 'EC 95862' (Chandan *et al.*, 2008).
- 'Langra' grafted on 'Bappakai' rootstock recorded the highest fruits number/plant followed by 'Vellaikulumban' and 'Chandrakaran' (Singh and Kanpure, 2006).

- Rootstock trial conducted at IIHR, Bangalore for 21 years, indicated no significant effect of rootstocks (Vellaikulumban, Bappakai, Chandrakaran, Kurukan, Muvandan, Mylepelian and Olour) on fruit quality of 'Alphonso' mango (Reddy *et al.*, 2003).
- Screening of mango rootstocks to salinity has shown that the polyembryonic cultivars 'Olour' and 'Bappakai' could withstand higher level of salinity (Palaniappan, 2002).
- (Dubey *et al.*, 2007) reported that 'Olour' as best salt tolerant rootstock compared to 'Kurukan'. Mango cultivar '13-1' was selected as a polyembryonic (3-6 embryos) rootstock for calcareous soils or for irrigation with saline water.
- Mango trees on '13-1' rootstocks showed excellent performance on soil containing 20% lime, three other cultivars on '13-1' rootstock showed good development on sandy soil with 10-20% lime (Gazit and Kadman, 1980).
- (Zuazo, 2004) indicated the greater capability of 'Gomera-1' rootstock to saline conditions.
- Seedlings from stone of 'Kesar' variety was found to be better with significantly highest survival percentage, germination percentage and growth parameters with higher salt concentrations of water (Varu and Barad, 2010).

3.1.2 Role of rootstocks in citrus

Role of rootstocks in citrus is one of the most debatable and discussed issues and its selection is a major consideration under planning of any citrus orchard. One thing can be safely said that choosing the right rootstocks is fundamental to the success of the orchard. The characteristics of some selected citrus rootstocks are given in following Table:

Root Stock	Horticultural performance			Reaction to						
	Yield	Quality	Vigour	Root rot	Nematode	Tristeza	Exocortis	Salt	Drought	Root
Rangpur lime	G	M	G	MT	S	R	S	R	R	D
Marmalade orange	G	M	G	MT	MT	R	S	R	R	D
Rough Lemon	G	L	G	S	S	R	R	T	T	D
Cleopatra Mandarin	M	M	M	T	S	R	R	MT	S	M
Sour orange	G	G	M	R	T	HS	T	T	MT	D
Sweet lime	M	M	G	S	MT	S	S	S	S	M
Trifoliate citrange	L	M	L	R	R	R	H	HS	HS	SH

G = good, M = Moderate, L = Poor or low, R = Resistant, T = Tolerant, MT = Moderately tolerant, S = Susceptible, HS = Highly susceptible, D = Deep, M = Medium, SH = Shallow.

3.1.3 Role of rootstocks in grapes

- Although grapevines can grow in soils with a wide range of pH (4.5-6.5), very acid soils present a problem. The use of acid-tolerant rootstocks, such as 'SO4' and '3309C', was highly recommended.
- According to (Leon *et al.*, 1969) grape rootstocks *V. champini* and *V. vinifera* are considered to be tolerant to salinity.
- Similarly (Kocksis *et al.*, 1999) suggested that hybrids of *V. berlandieri* x *V. riparia* were more tolerant to drought.

3.1.4 Role of rootstocks in guava

- At present, guava is propagated on seedlings raised from open pollinated seeds and not on uniform clonal rootstocks.
- (Edwar and Shankar, 1964) reported that *P. cattleianum, P. guinesee, P. molle* and Philippine guava were found suitable as rootstocks.
- On *P. cattleianum,* the trees were the tallest and gave the maximum yield.
- *P. pumilum* rootstock had dwarfing effect but fruits on this rootstock had maximum number of seeds, highest TSS and total sugars.
- Trees on *P. cujavillis* produced the largest fruits with the highest ascorbic acid content, though they were rough-skinned and not uniform.
- (Sharma *et al.*, 1992) recorded that rootstock aneuploid No. 82 impart dwarfness to 'Allahabad Safeda' in terms of plant height, plant spread and tree volume. They also recorded higher yield on this rootstock.

3.1.5 Role of rootstocks in apple

- There is a great diversity in the type of material used for raising rootstock seedlings in apple.
- In Kashmir, wild indigenous 'Crab apple' known as 'Trel' is used, in Himachal Pradesh, seedlings of 'Crab C' are employed as rootstock.
- The first rootstock trial of apple was initiated in 1937 at Chaubattia with 'Red Delicious', 'Jonathan' and 'Rymer' cultivars using Crab C, M 2, M 13, Merton 779, Merton 793 and 'Local Selection' (seedling selection from 'Ribbistin Pippin') rootstocks.
- After the introduction of M and MM series of rootstocks during early sixties, from East Malling Research Station, England, at Mashobra and Kotkhai in HP, Chaubattia in UP and Shalimar in J & K, elaborate varietal-cum-rootstock trials were initiated at Mashobra in 1967, Kotkhai in 1968, Chaubattia in 1969 and several locations in J & K in 1969.

- As a result, some promising rootstocks were identified for different apple growing regions in India *viz.*
    - M 7, M 9, M 26, MM 106 and MM 111 for Himachal Pradesh;
    - M 2, M 4, M 7, M 9 for J&K and Merton 779, MM 106, M 13 for Uttarakhand

3.1.6 Role of rootstocks in pear

- Fewer rootstocks choices are available for pears than for apples.
- Domestic pear seedlings (*Pyrus communis*) are still the most acceptable rootstocks for pear cultivars in terms of vigour, hardiness, and compatibility.
- However, all pear trees on seedling roots are susceptible to fire blight.
- Seedlings of *Pyrus calleryana* are adapted to many soil conditions and produce semi-vigorous growth. It is resistant to fire blight, but it's not sufficiently winter hardy.
- *Pyrus betulaefolia* seedlings are also adapted for many soil conditions and it's a winter hardy. The trees are vigorous, larger than 'Bartlett seedling', and moderately tolerant to fire blight.
- *Pyrus ussurensis* can be used as resistant to fire blight, pear psylla, and cold hardy.

3.1.7 Role of rootstocks in stone fruits

- In India, the stone fruits like peach, plum, apricot and almond are generally propagated on their own seedling, while the cherry plants are raised on 'Paja' (*Prunus cerasoides*) seedlings.
- Commonly 'Behmi (*P. mira*)' is widely used as a rootstock for almond.
- Peach seedlings generally shows susceptibility to nematode but 'Nemaguard' and 'Okinawa' showed tolerance to nematode attack.
- Some nurseries also use 'Behmi' (*P. mira*) as rootstock for peach.
- For Plum (*P. domestica*) selections like 'Brompton' and 'Common plum' from; St. Julien, 'Common Mussel' and 'Damson' from *P. institia*; Myrobalan from *P. cerasifera*; peach (*P. persica*), apricot (*P. armeniaca*), almond (*P. amygdalus*) and some hybrids like Marianna plum (*P. cerasifera* × *P. munsoniana*) are used as rootstocks in other countries.
- In India, peach, apricot, behmi (natural hybrid of almond and wild peach) and plum seedlings are commonly used as rootstocks (Singh and Gupta, 1971).

- In India, the most common rootstock for cherry in Kashmir, Himachal Pradesh and Kumaon Hills of UP is 'Paja' (*P. cerasoides*), though old plantations are invariably on Mazzard or Mahaleb rootstocks.

3.2 Use of root stocks in fruit crops

Sr. No.	Fruits	Root stocks	Specific feature
1.	Almond	Bitter almond, almond seedling, Mariana 2624, Nemaguard, Sloh	Sloh (Peach × Almond), Self fertile
2.	Apple	**Seedling rootstock**: Crab apple (*Malus baccata*) **Clonal rootstock:**	
		M-9	Dwarf, suitable for HDP
		M-27	Ultra-dwarf, suitable for HDP
		MM 104 & MM 106	Resistance to wooly apple aphid
		MM 111, MM 104	Resistant to drought
		MM 111	Vigorous
3.	Pear	**Seedling rootstock**	Kainth (*P. pashia*) commercially used in India
		Clonal rootstock EMLA Quince C	Ultra dwarf, suitable for HDP
		Old home, Farmindale	Resistant to fire blight
4.	Peach	Quince	It is widely accepted dwarf rootstock of pear. It is resistant to wolly aphids, nematodes, susceptible to oak root fungus, fire blight
		Quince A	Vigorous
		Quince B	Semi-vigorous
		Nemaguard, Nemared, Floridaguard	Resistant to Nematode
		GF 667 (Natural peach almond hybrid)	Tolerant to calcareous soils
		Titan hybrid	Highly vigorous, root-knot nematode resistance, tolerant to calcareous soils
5.	Plum	**Seedling rootstock**	Wild apricot
		Clonal rootstock	
		Myrobalan B	Resistant to cold, collar root and nematodes
		Mariana	Resistant to nematodes
		St. Julien	Most versatile rootstock for plums
		St. Julien A	Vigorous

		Pixy	Ultra dwarf rootstock
6.	Cherry	Gisela	Cold hardy
		Mazzard (*P. avium*)	Dwarf, commercial rootstock of cherry
		Colt	Semi dwarf, resistance to phytophthora, bacterial canker.
		Mahaleb	Most drought tolerant, suited to deep soils
7.	Apricot	Wild apricot	Commercially used in India
8.	Walnut	*J. regia* (Persian walnut)	Commonly used rootstock, it is unsuitable for waterlogged and saline soils.
		J. hindsii	Resistant to oak root fungus, root knot nematode but susceptible to phytophthora
		J. nigra	Tolerant to crown gall, phytophthora
		Paradox (*J. hindsi* × *J. regia*)	Tolerant to nematodes and phytophthora.
9.	Mango	Polyembryony rootstocks	Sabre, Chandrakaran, Bappakai, Olour, Goa, Salem, Kurukan, Nileshwar dwarf, Ambalavi, Muvandan, Nekkare, Peach
		Kurukkan	Salt tolerant
		Muvandan, Olour	Vigorous
		Vellaikolamban	Dwarf root stock for alphonso and dashehari
		13/1	Tolerant to calcareous soil
		Bappakai	Best for Neelum variety
		Gomera	Salt tolerant
		ST-9	Vigorous
10.	Citrus	Sour orange	Extensively used for sweet orange, mandarins, grapefruit and lemon. Adaptable to heavy soils
		Trifoliate orange	Dwarf rootstock, adapted to heavy soils. Tolerant to nematodes and tristeza but susceptible exocortis virus
		Rangpur lime	Suitable for Sathgudi Mosambi oranges, Nagpur and Kinnow mandarins. Adaptable to heavy soils, susceptible to exocortis and burrowing nematode.

		Rough lemon	It is widely used rootstock for sweet orange. It is tolerant to tristeza xyloporosis, exocortis and gummosis.
		Flying dragon	Ultra dwarf root stock
		Troyer citrange	Most promising rootstock for Nagpur mandarin and Kinnow mandarin in India
11.	Guava	Pusa srijan (aneuploid tetrasomic, 2n+2)	Potential for dwarfing, resistant to wilt
		P. friedrichsthalianum	Dwarf, resistant to wilt, nematode
		P. molle	Resistant to guava wilt
		P. pumilum	Highly dwarf, high sugar content
		P. cujavilis	Produce largest but non-uniform and rough skinned fruits
12.	Ber	*Z. nummularia*	Dwarf, salt tolerant.
		Z. rotundifolia	Tolerant to drought
13.	Fig	*Ficus glomerata*	Resistant to Root Knot Nematode
14.	Custard apple	*Annona glabra:* Pond apple	Tolerant to flood condition
15.	Jamun	*Syzygium densiflora*	Resistant to termite
		Syzygium fruticosum	Grown as wind break
16.	Sapota	Rayan or Khirni *(Manilkara hexandra)*	Commercial rootstock in India. Wider soil adaptability
17.	Grapes	Salt creek	Salt and nematode resistant
		Temple	Multiple resistant to Pierce's disease
		St. George, riparia and gloria	Phylloxera resistant
		Dogridge	Suitable for drought and salinity prone areas

References

Chandan, P. M.; Kadam, J. H. and Ambad, S. N. (2008). Effect of different polyembryonic and monoembryonic rootstocks on performance of Dashehari mango. Inst. J. Agril. Sci., 2(2): 594-595.

Dubey, A. K.; Srivastav, M.; Sharma, Y. K.; Pandey, R. N. and Deshmukh, P. S. (2007). Dry mass production and distribution of nutrients in two mango rootstocks as affected by salinity. Indian. J. Hortic. 64: 385-390.

Edwar, J. C. and Shankar, G. (1964). Rootstock trial for guava. Allahabad farmer, 38:249-50.

Gazit, S. and Kadman, A. (1980). 13-1 mango rootstock selection. Journal of Horticultural Science, 15(5):669.

Jauhari, O. S.; Teaotia, S. S. and Upadhyay, S. K. (1977). Acta Horti., 24: 107-109

Leon, B.; Ehlig, C. F. and Clark, R. A. (1969). Effects of grape rootstocks on chloride accumulation in leaves. J. Amer. Soc. Hort. Sci., 94: 584-590.

Palaniappan (2002). Germplasm screening for salinity stress in tropical fruit species. Regional Training Course "Characterization, Evaluation and Conservation of Tropical Fruits Genetic Resources", organized by IPGRI, ICAR and IIHR.

Quijada, O.; Castellano, G.; Jiménez, N. and Briceno, E. (2008). Development and productivity of mango cultivars grafted over sinamaica rootstcks in the plain of Maracaibo. Agronomía Trop.,58(3): 207-213.

Reddy, Y. T. N.; Reju, M.; Kurian, P. R.; Ramachander, G. S. and Kohli, R. R. (2003). Long term effects of rootstocks on growth, fruit yielding patterns of Alphonso mango (M. Indica) Scientia Hort., 97(2): 95-108.

Sharma, Y. K.; Goswami, A. M.and Sharma, R. R. (1992). Effect of dwarfing aneuploid guava rootstock in high density orcharding. Indian J. Hort., 49: 31-36.

Singh, R. N. and Guptha, P. N. (1971). Rootstock problem in stone fruits and potentialities of wild species found in India. Punjab Hort. J.,11: 157-175.

Varu, D. K. and Barad, A. V. (2010). Standardization of mango rootstock for mitigating salt stress. Indian J. Hort.,67: 79-83.

4

Planting System and High-Density Planting

Mithapara Kinnari D., Vrutti Patel and V. M. Parmar

4.1 Planting systems

4.1.1 Vertical row planting pattern

1. **Square system:** In this system, trees are planted on each comer of a square whatever may be the planting distance. This is the most commonly followed system and is very easy to layout. The central place between four trees may be advantageously used to raise short lived filler trees. This system permits inter cropping and cultivation in two directions.

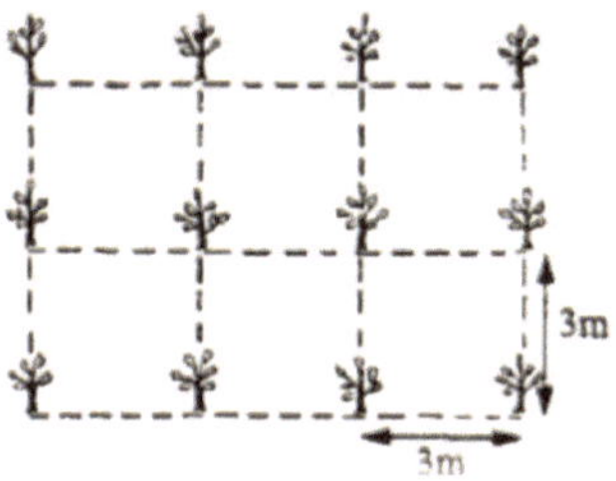

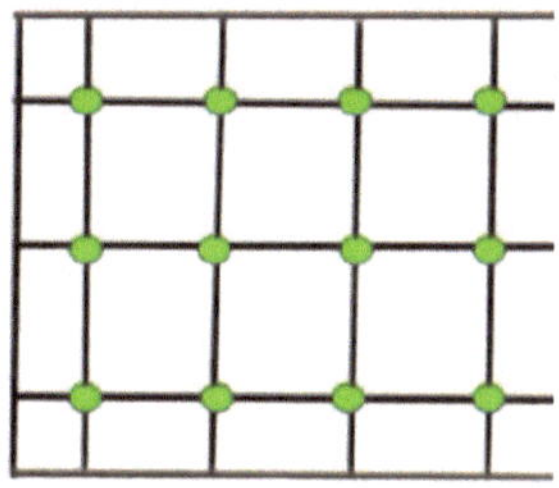

2. **Rectangular system:** In this system, trees are planted on each corner of a rectangle. As the distance between any two rows is more than the distance between any two trees in a row, there is no equal distribution of space per tree. The wider alley spaces available between rows of trees permit easy intercultural operations and even the use of mechanical operations.

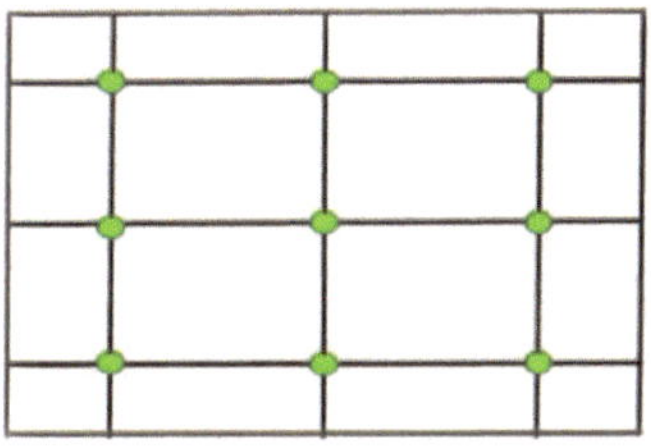

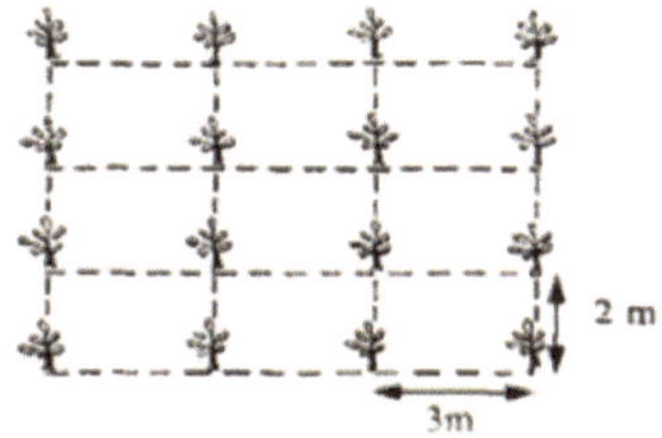

4.1.2 Alternate row planting pattern

3. Hexagonal system

- In this method, the trees are planted in each comer of an equilateral triangle.
- This way six trees form a hexagon with the seventh tree in the centre.
- Therefore, this system is also called as 'septule' as a seventh tree is accommodated in the centre of hexagon.
- This system provides equal spacing but it is difficult to layout.
- The perpendicular distance between any two adjacent rows is equal to the product of 0.866 x the distance between any two trees.
- This system accommodates 15% more trees than the square system. The limitations of this system are that it is difficult to layout and the cultivation is not so easily done as in the square system.

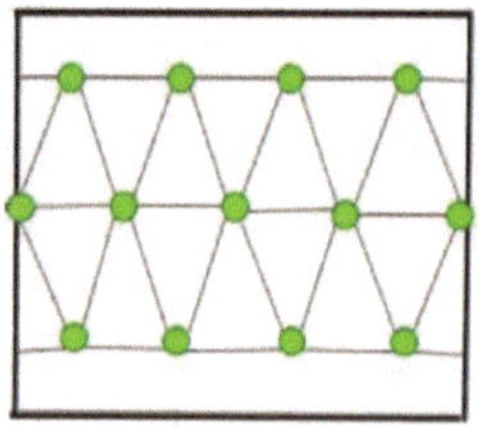

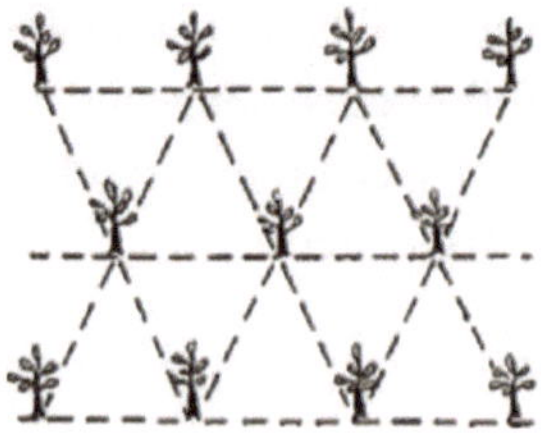

4. Diagonal or Quincunx system

- This is the square method but with one more plant in the centre of the square.
- This accommodates double the number of plants, but does not provide equal spacing.
- The central (filler) tree chosen may be a short lived one.
- This system can be followed when the distance between the permanent trees is more than 10m.
- As there will be competition between permanent and filler trees, the filler trees should be removed after a few years when main trees come to bearing.

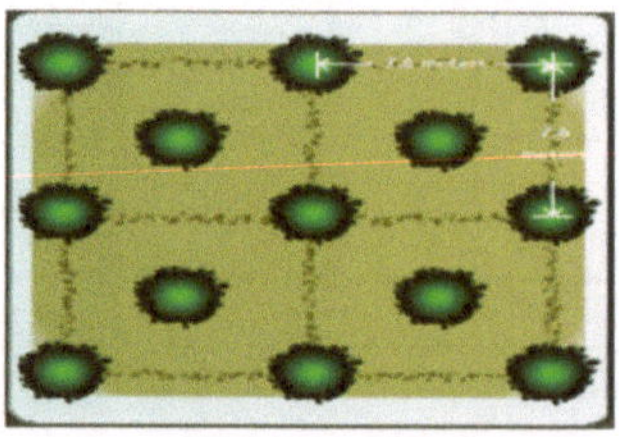

5. **Triangular system**
   - The trees are planted as in square system but the difference being that those in the even numbered rows are midway between those in the odd rows instead of opposite to them.
   - Triangular system is based on the principle of isolateral triangle.
   - The distance between any two adjacent trees in a row is equal to the perpendicular distance between any two adjacent rows.
   - However, the vertical distance, between immediate two trees in the adjacent rows, is equal to the product of (1.118 x distance between two trees in a row).
   - When compared to square system, each tree occupies more area and hence it accommodates few trees per hectare than the square system.
6. **Contour system**
   - It is generally followed on the hills where the plants are planted along the contour across the slope.
   - It particularly suits to land with undulated topography, where there is greater danger of erosion and irrigation of the orchard is difficult.
   - The main purpose of this system is to minimize land erosion and to conserve soil moisture so as to make the slope fit for growing fruits and plantation crops.
   - The contour line is so designed and graded in such a way that the flow of water in the irrigation channel becomes slow and thus finds time to penetrate into the, soil without causing erosion.

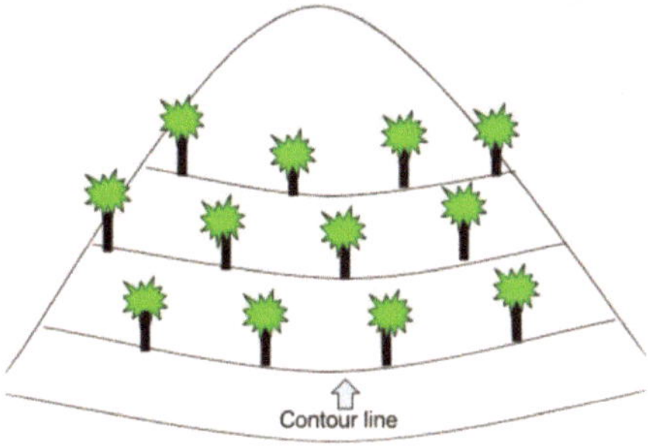

7. **Terrace system**
   - Terrace system on the other hand refers to planting in flat strip of land formed across a sloping side of a hill, lying level along the contours.
   - Terraced fields rise in steps one above the other and help to bring more area into productive use and also to prevent soil erosion.

- The width of the contour terrace varies according to the nature of the slope.
- If the slope becomes stiff, the width of terrace is narrower and vice-versa. The planting distance under the contour system may not be uniform.

4.1.3 Planting systems: Banana

8. Paired row system

- In this method, distance among one single pair of plants is less but between pairs is more.
- The most widely followed paired row system spcing is: 1.2 m × 1.2 m × 2.0 m, accommodates 5208 plants/ha.

4.1.4 Recommended spacing of various fruit crops

Sr. No.	Crop	Traditional Spacing (m)
1.	Mango	10 × 10
2.	Banana	1.8 × 1.8
3.	Papaya	1.8-3 × 1.8-3
4.	Guava	7 × 7
5.	Sapota	10 × 10
6.	Jack fruit	10 × 10
7.	Litchi	10 × 10
8.	Custard apple	4.5 × 4.5
9.	Pomegranate	5 × 5
10.	Ber	8 × 8

4.2 High-Density planting

- The high-density planting (HDP) in fruit crops is one of the recent novel concepts of increasing the productivity without affecting the quality of fruits.
- India is the largest producer of fruits in the world after China.
- The average productivity and per capita availability of fruits in India is, however, low as compared to many developed countries.
- The main reasons for low productivity are old and senile orchards, wider spacing, low yielding varieties, poor orchard management and inadequate technological up-gradation and adoption by the growers.
- Presently, the continuing decline in the availability of cultivable land, rising energy and land costs together with the increased demand of fruit and fruit products, have given thrust to the concept of high-density planting (HDP) in fruit crops.
- High density planting gives earlier production and return per unit area, shortens juvenility, eco-friendly, provides efficient land use and better use of resources like light, water and fertilizers, efficient pesticides application, besides, in this system the harvesting becomes easy.
- The success of this technology in most of the fruit crops is dependent on the use of methods to control shoot growth and maximize light interception as the trees begin to bear fruit (Menzel and Lagadec, 2014).
- The conceptual background of high-density planting in fruit growing was pioneered in temperate fruits and first planted at the end of the nineteen sixties, since then there is rise in establishment of commercial high density orchards throughout the world.
- High density planting in fruit trees was first introduced in case of apple in Europe during the nineteen sixties, with the development of Malling and Malling - Merton rootstocks.
- Subsequently, this was experimented in other temperate fruit crops like peach, plum, pear and sweet cherry. Today, majority of new orchards of apple and to some extent that of other fruit crops in Europe, New Zealand, USA and Australia use this intensive system of fruit production profitably Iyer and Kurian (2006).
- In India, high density plantings have successfully been demonstrated in
    - guava (Lal *et al.*, 2007),
    - litchi (Mishra *et al.*,2014),
    - mango and papaya (Ram, 1996).

- **Advantages of high-density planting**
  - It induces the precocity
  - Enhanced fruit yield and quality
  - Low cost per unit production
  - Enables mechanization in fruit crops
  - Efficient use of applied and natural resources
  - Harvesting is easy and cost of picking is easy
  - The ratio of fruiting shoots to supporting one is higher
  - Ease in spraying of chemicals for pest and disease control
- **Disadvantages ofhigh-density planting**
  - In long run, it results in heavy competition for space, nutrients and water.
  - Overcrowded growth of canopy results in building up of high humidity, reduced ventilation in the orchard, which are conducive for more incidence of disease pest.
  - The trees in close plantings soon begin to crowd each other and yields decline.
  - Efforts to breed dwarfing rootstocks that can control the growth of mature trees have been largely unsuccessful.
  - Effective canopy management appears to be the largest barrier to success of high density orchards.
  - For restriction of growth in high density orchards, growth retardants are used. There are reports that some of the growth regulators persisting in the harvested fruit and soil (Menzel and Lagadec, 2014).
  - Reduced fruit size and quality.
- **Approaches of high-density planting**
  - HDP can be achieved with the suitable use of following components, they are :

  a) Dwarf scion varieties,

  b) Dwarfing rootstocks and inter-stocks,

  c) Training and pruning,

  d) Use of growth regulators and

  e) Suitable crop management practices.

  - These components are harnessed in HDP which helps in attaining the goal of high yield and quality.

a) **Use of dwarf scion varieties**

- It is easier to establish high density orchards if the trees are naturally small. Use of genetically dwarf cultivar offers great scope for close plantings, which have the potential for higher yields and returns than traditional plantings.
- However, availability of dwarfing scion cultivars are meager as indicated below:

Crop	Dwarf cultivar
Banana	Dwarf cavendish, Jahaji, Grand Naine, Poovan, Rasthali, Ney Poovan
Guava	Pant prabhat, Red Fleshed
Litchi	Culcuttia, China
Mango	Amrapali, Arunika, ArkaAruna,
Papaya	Pusa dwarf, PusaNanha
Sapota	PKM-1, PKM-3

b) **Use of dwarfing root stocks**

- Root stocks are known to have a profound effect on the tree vigour, precocity, productivity, quality of fruits and longevity of varieties grafted on them.
- Dwarfing can be due to the rootstock or the scion, or both.
- Attempts have made to standardize dwarfing rootstocks especially in the fruit crops like ber, citrus and grape. However, much choice is not available for most of the tropical and subtropical fruit crops in comparison to the temperate fruits.
- Use of graft incompatible scion and rootstock induces dwarfness in the composite plant.
- As in case ber, if scion cultivars grafted on to the *Zizyphusrotundifolia*, *Z. nummularia*, dwarfness may be induced due to graft incompatibility.

Crop	Dwarf rootstocks
Ber	*Zizyphusrotundifolia, Z. nummularia*
Citrus	Troyer citrange, Flying Dragon (*Poncirustrifoliata),* KarnaKhatta
Guava	*Psidiumfriedrchsthalianum, P.chinensis,* PusaSrijan, Aneuploid-82
Mango	Vellaikullamban, Totapuri red, Rumani, Olour

c) **Use of pruning to control tree growth**

- Overcrowding poses a serious problem for orchard access and for adequate light interception needed for optimum photosynthesis, flowering and fruit set and quality.

- Pruning and production of new shoot is found essential to maintain continuity of vigour and to provide bud for flower emergence.
- Pruning can be used to increase the distribution of light through the canopy, although productivity is not always restored in the same season.

d) **Use of growth regulators to control tree growth**

- Pruning often leads to strong re-growth of shoots in mango and other fruit crops.
- Growth retardants are plant growth regulators (PGR) that retard growth of plants. There are mainly three mechanisms by which tree height can be restricted by chemicals:

1. Severely inhibiting apical meristemic activity
2. **By reduced apical control**: Reduction of tree height may also result from simultaneous growth of many shoot axes, which owing to competition for limited nutrients reduce the elongation of one or more axes on the same root system. Promising chemicals for inducing auxiliary bud growth include TIBA (2,3,5- Triiodobenzoic acid), which is an inhibitor of auxin synthesis and translocation, cytokinins, particularly benzyl adenine, which promotes the growth of auxillary buds.
3. **By inhibiting internodes elongation without disrupting apical meristamic functions**

- Specific inhibition of internodes elongation without concomitant disruption of apical meristamatic function and loss of apical controls caused by certain growth retardants like, SADH (Succinic acid-2-2 dimethyl hydrazide), CCC, paclobutrazol.
- These growth regulators reduce gibberellins's biosynthesis, though different retardants act at different steps in biosynthesis pathway.
- SADH prevents the conversion of trans-geranyl pyrophosphate to Kaurenoic acid from getting converted to GA, whereas Paclobutrazol retards formation of Kaurene-19-ol from Kaurene. The chemical when applied to the leaves is readily translocated through phloem to the apical meristem, where it results in temporary check in apical growth, thereby promoting sprouting of auxiliary buds and internodal elongation.

- **Meadow orcharding: Guava**
  - UHDP or meadow orchard system is a new concept of planting which has been developed in guava for the first time in India at CISH, Lucknowby Dr. Gorakh Singh.
  - The Meadow Orchard is a modem method of fruit cultivation using dwarf tree with modified canopy.
  - Better light distribution within tree canopy increases the number of well illuminated leaves.
  - It also promotes rate of photosynthesis that leads to high yield per unit area.
  - This system of guava planting is going to revolutionize the guava industry by enhancing productivity coupled with reduction in production costs.
  - The meadow orchard system of guava accommodates 5000 plants ha^{-1}, planted at 2.0 x 1.0m spacing and managed with regular topping and hedging especially during initial stages.
  - Topping and hedging in guava are helpful in controlling tree size and extending fruit availability.
  - Standard spacing for guava is 6x 6m, which accommodates277 plants ha^{-1}.
  - It is important to shape the tree after planting. Spacing of the plants in the system depends on fertility of the soil, availability of water, intensity of sunlight and wind exposure.
  - As guava responds very well to pruning, the following plant densities have been recommended by Central Institute for Subtropical Horticulture, Lucknow for early, higher and quality guava production:

A. 3.0m(row to row) x 1.5m(plant to plant) accommodating 2222 plants/ha

B. 3.0 m (row to row) x 3.0 m (plant to plant) accommodating 1111 plants/ha

C. 6.0 m (row to row) x 3.0 m (plant to plant) accommodating 555 plants/ha

Comparison between traditional and meadow orchard system of guava growing

Attributes	Traditional system	Meadow system
Bearing	After two years	From first year
Production	Average yield is 12-20 t ha^{-1}	Average yield is 40-60 t ha^{-1}
Management	Difficult to manage due to large tree size	Easy to manage due to small tree size
Labour requirement	Requires more labour	Requires less labour
Production cost	Higher cost of production	Lower cost of production
Harvevsting	Difficult	Easy
Quality	Large canopy, poor sunlight penetration and poor quality fruits.	Small canopy, better air and sunlight penetration, minimum disease incidence and high quality fruits with good color development

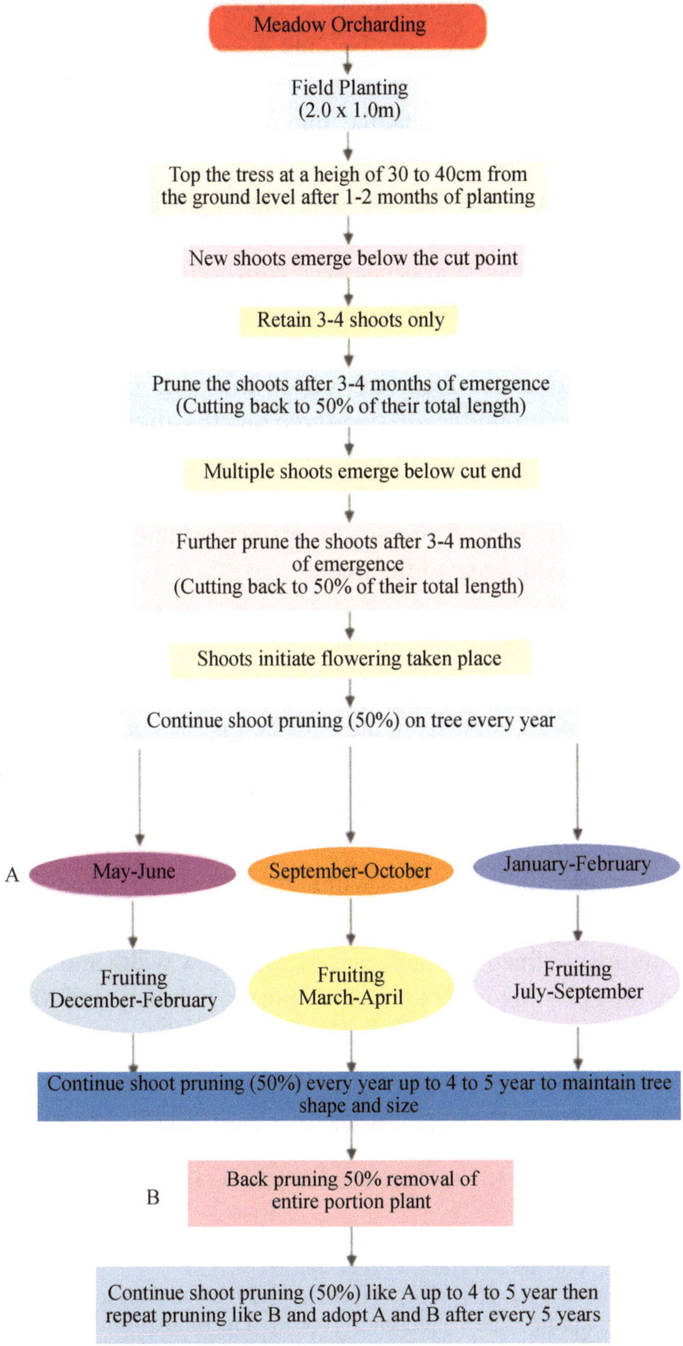

(***Source:*** CISH, Lucknow)

- **Canopy management for guava in meadow orcharding**
  - Untrained or unpruned guava trees become huge and unmanageable after a couple years of growth.
  - The bearing area is reduced and the interior ofthe plant becomes entirely devoid offruiting.
  - Trees are pruned to increase the yield of quality fruits by eliminating crowded and criss-cross branches. Pruning begins at an early stage of plant growth to develop single trunk trees with well-spaced scaffold branches to form the frame work.
  - Apical growth is to be controlled within the first year of planting for better canopy architecture. Trees are topped to a uniform height of 60 - 70 cm from the ground level, two-three months after planting to induce the emergence of new growth below the cut point. As a result, new shoots emerge.
  - Three to four equally spaced shoots are retained around the stem to form the main scaffold limbs of the tree.
  - These shoots are allowed to grow for4-5 months after topping until they attain a length of about 40 - 50 cm.
  - The selected shoots are further pruned to 50 per cent of their length for inducing multiple shoots from the buds below the cut end.
  - Newly emerged shoots are allowed to grow up to 40-50 cm and are pruned again for the emergence of new shoots. This is mainly done to obtain desired shape.
  - After two years, the short branches within the tree canopy produce a compact and strong structure.
  - All the plants are confined to a hedge shape of2m row width and 2.5m height for which pruning is performed in January-February and May-June every year.

References

Iyer, C. P. and Kurian, R. M. (2006). High Density Planting in Tropical Fruits – Principles and Practises. International Book Distributing Co., Lucknow, India.

Lal, S.; Tiwari, J. P. and Mahajan, A. R. (2007). Studies on planting systems in guava cv. Sardar. Acta Hort., 735: 263-266.

Menzel, C. M. and Lagdec, M. D. (2014). Increasing the productivity of avocado orchards using high density plantings: A review.Sci Hort., 177: 21-36.

Mishra, D. S; Lal, R. L. and Chand, S. (2014). Precision litchi culture technologies in Uttarakhand.In:Recent Plasticulture Approaches Towards Precision Horticulture, pp. 73-77.

5

Crop Modelling

D.R. Kanzaria, F.P. Karagatiya and H.J. Senjaliya

5.1 Crop modelling

- **Crop model**
  - "Schematic representation of the system''*i.e.* a model as defined by De Wit 1970.
  - Such a model can be based on statistics applied to a set of experimental data or on physical laws (Guyot, 1997).*e.g.* Irrigation is able to satisfy a crop's water needs provided its management relies on the measurement or estimation of evapotranspiration. The manager can either invest in equipment to monitor changes in water content in the soil or fluxes of water into the plant or the atmosphere, or rely on a ''schematic representation of the system'', *i.e.* a *model.*
  - In very simple terms we can define a crop model as "a well-thought-out plan of activities to grow the desired crop to get maximum possible yields of high-quality produce with the given resources. It can be a mental or a written exercise (a "schematic representation of the system") that a farmer or a farm manager prepares for a crop or for an individual crop".
  - In other words, we can also say that a model is "an attempt to describe a certain process or system through the use of a simplified representation, preferably a quantitative mathematical expression, that focuses on a relatively a few key variables that control the process or system".
  - It is clear that in finalizing a crop model we have to assume some mean or standard values (based on the past record, experience, expert judgment, *etc.*) of the variables which we may not be taking explicitly in our model.
  - Since, there can be a large number of factors that may affect the crop, it becomes easier to understand their effect if these variables are grouped on some basis. Thus, we may have a group of ecological

variables, economic and social variables, technical and climatic factors that may need to be considered in our crop models.

5.2 Utility of crop model

- Crop models help us to understand the crop production process or system in a more systematic way.
- The process of modeling is often equated to solving of a puzzle. A puzzle has to be considered as a whole even if we need to fit a one small block. Crop models provide us quantitative information about the number of inputs like the doses of fertilizers, number of irrigations, amount of insecticides/pesticides, *etc.* required.
- These models also help us to consider various input requirements under different climatic conditions. This means models help us to get reasonably clearer picture of otherwise hazy scene.

5.3 Types of models

- Based on different groups of variables that affect crops production, we can categories crop model or models as:
  1. Ecological models
  2. Climatic models
  3. Socio-economic models
- The climate-based models are formulated to predict the correlation of climatic variables (like temperature, rainfall or snowfall, humidity, day length, chilling hours, etc.) with a given biological phenomenon like survival rate of species, growth of vegetative parts, fruit-set, crop maturity or the crop yields, *etc.* These models are also used by the ecologists to identify the variables which affect the crop yields through their impact on modifying the local climate.
- Understanding the interactions of farming systems with the surrounding environment is thus important for success in agribusiness and modelling helps us in doing the same.
- The crop models with as many variables as possible are thus better in making predictions than the models with a few variables. However, modelling with a very large number of variables may not always be possible because of the problems of lack of relevant data, mathematical programming problems, *etc.* Most of the times, therefore, we have models with the variables in which we have the immediate interest.

5.4 Production function – an algebraic representation of a model

- Most common crop models are the economic models analyzing the effect of some inputs on crop yields.
- In these models, influence of an independent variable is estimated on the crop yield or on the growth of the crop.
- We know that a production process is a set of sequence of rules for using different inputs that needs to be followed for using different inputs.
- The ultimate outcome of a production process is the final output of that process, i.e. the crop yield or the desired vegetative growth.
- In a simple modeling scenario, a production process can be considered as a production function (a mathematical or the symbolic expression of a production process). A production function is the most common crop model. In general, a production function can be algebraically represented as;
- Y = f (X),
- Where; Y is the dependent variable (say output) which depends on the amount of input (X) used.
- Since we can change the level of input being used, X, is also called an independent variable. We assume that the level of technology being used remains the same during the production process.
- For a multiple inputs case, a production function will take the following form; indicating that the final output depends on, or is a function of 'n' inputs or 'n' independent variables.
- Y = f (X1, X2, X3, . . . , Xn-1, Xn)
- However, it is common to analyze the effect of a single factor of production on an output; while the other variables are held constant at their mean use level or at predetermined levels. We can thus rewrite the above equation for a production function as follows;
- Y = f (X1/ X2, X3, . . ., Xn-1, Xn)
- We can graphically represent the response of a dose of an independent variable, or a package of inputs, on the dependent variable (output) by plotting the output obtained by the use of each dose of the input in a production trial. Such a representation of the production data is termed as the response curve or the total product curve.
- The shape of the response curve for a product depends upon the nature of the product and the production process being used. The response of different inputs, under given production conditions may be different over the range of a response curve.

- Average response, or average product (AP) as it is commonly known, is defined as the output produced per unit of input used. Marginal product (MP) or marginal response, on the other hand, is defined as the change in total output resulting from a unit increase in input.
- It is possible that for an activity we might notice increasing, constant and decreasing returns over the range of total product or response curve.
- This happens as a result of varying capacity utilization of the fixed resource due to increased use of variable inputs.
- As for example, in case of crops, unit response to input use is higher in the initial stages when the productive capacity of a fixed resource, say for example land, is not fully utilized. So, when more input is used the output obtained per unit of additional input used is increasing.
- Later, the response may be constant over a restricted range and then it declines when the land resource gets over utilized.
- Most common examples of such responses are the fertilizer applications on a given piece of land or the number of irrigations in an irrigation trial.
- In both these cases, the yield response to initial input doses are higher and the response declines as we continue using additional doses of inputs. Finally, there may be losses due to toxicity or lodging *etc.*, if the input use is continued beyond the absorptive capacity of land and the crop.
- Thus, it becomes important to decide as to what level of production (and thus the level of input use) is desirable or profitable and where one should stop using the input.
- It is important to consider economic aspects (profitability) because the inputs are scarce and have to be paid for.
- If the inputs are not scarce or are not paid for, then in general, one might target the biologically maximum yield.
- This mostly is the case in research trials where we need to find out the maximum yield or total production of the crop.

5.5 Application in horticulture

- Modeling has become relatively easy now with the use of computers.
- Computers help in reducing the complications of handling voluminous data.
- Therefore, apart from annual crops, modelling for perennial fruit trees is also commonly done.

- In the field of horticulture, the crop models are needed for modelling yield forecast, policy analysis and management options.
- It is very important to predict yields so as to prepare ourselves for marketing of the produce in a hassle-free manner.
- For example, it is important to know the timings of the arrivals of different produce in the market. If, through modelling, we can get an idea when the marketing season would be a lean season, and how we can change our harvesting to target this period, we can increase our profits.
- In perennial crops/fruit trees we need to model "the architecture of fruit trees and the relations between pruning and flower and fruit development" for maintaining the vigour of the trees.
- Such modelling is generally called biological modelling.
- When we combine the biological and economic considerations, we say we are doing a bio-economic modelling.
- The economic content of the model relates to the costs and returns from the production process while the biological content refers to the physiological production processes from planting to maturity of the crop.
- The annual fruit production is determined by the biophysical model which may use factors like tree density, pruning/training regimes.
- As a branch of science progresses from the qualitative to the quantitative, one day it may be expected to reach the point where the connections between theory and experiment are most efficiently made using the language of mathematics (Thornley and Johnson, 1990).
- Significant advances in modelling leaf and canopy photosynthesis come from studies on tomato and chrysanthemum plants (Acock, 1991).
- The concepts of growth analysis have extensively been worked out on lettuce and tomato crops (Hunt *et al*., 1984; Thornley and Hurd, 1974).
- Fruit trees and part of the ornamental species are perennials; original modelling studies have been carried out to simulate specific phases of their development such as dormancy (Erez *et al*., 1990).
- Most horticultural species present discontinuous canopies with complex consequences on light interception, aerodynamics and gas exchange.
- This has justified coupled approaches integrating micrometeorology and crop physiology (Daudet and Tchamitchian, 1993; Gijzen and Goudriaan, 1989).
- Greenhouse crops are grown in conditions that may differ significantly from natural conditions.

- It is not surprising that studies on the effects of high CO concentrations in the atmosphere 2 focused on greenhouse crops before the problem of the consequences of global climatic changes arose in the eighties (Enoch and Kimball, 1986).
- In all of these examples, the main objective has been the improvement of knowledge.
- It is clear, for example, that most of the scientific debate around the controls of photosynthesis from the chloroplast to the canopy has been organised around the design and testing of mathematical models (cf. the various contributions in Boote and Loomis, 1991).

5.6 Fruit tree models

- The perennial nature of fruit tree implies that tree growth and the yields are also dependent on previous years' growth and health of the tree.
- In deciduous fruit trees, for example, the number of flowers, and thus crop potential is determined by the condition of the plant in previous season.
- In fruit trees there is "a perennial woody skeleton which not only grows and develops during the annual cycle, but also accumulates and exports carbohydrates and other reserve materials. This complicates the estimation of the net increase in biomass of a fruit tree during the annual cycle".
- Nevertheless, we model the response by taking various assumptions or by keeping some of the factors at a known level in the model.

References

Acock, B. (1991). Modeling canopy photosynthesis response to carbon dioxide, light interception, temperature, and leaf traits. In: Boote, K.J., Loomis, R.S. Eds., Modeling Crop Photosynthesis—From Biochemistry to Canopy. Crop Science Society of America, Madison, USA, pp. 41–56.

Boote, K. J. and Loomis, R. S.(1991).Modeling Crop Photosynthesis—From Biochemistry to Canopy. Crop Science Society of America, Madison, USA, 140 pp.

Daudet, F. A. and Tchamitchian, M. (1993). Radiative exchange and photosynthesis. In: Varlet-Grancher, C., Bonhomme, R., Sinoquet, H. Eds., Crop Structure and Light Microclimate: Characterization and Applications. INRA Editions, Paris, pp. 401–417.

De Wit, C. T. (1970). Dynamic concepts in biology. In: Setlik, I. Ed., Prediction and Measurement of Photosynthetic Activity. Pudoc, Wageningen, pp. 17–23.

Enoch, H. Z.; Kimball, B. A.(1986). Carbon Dioxide Enrichment of Greenhouse Crops: 2. Physiology, Yield, and Economics. CRC Press, Boca Raton, FL, 230 pp.

Erez, A.; Fishman, S.; Linsley-Noakes, G. C. and Allan, P. (1990). The dynamic model for rest completion in peach buds. Acta Horticulturae,276: 165–174.

Gijzen, H. and Goudriaan, J. (1989). A flexible and explanatory model of light distribution and photosynthesis in row crops. Agric. For. Meteorol. 48: 1–20.

Guyot, G. (1997). Climatologie de l'Environnement. Masson, Paris, English edition in press, Praxis Editing, London, pp. 505.

Hunt, R.; Warren-Wilson, J.; Hand, D. W. and Sweeney, D. G. (1984). Integrated analysis of growth and light interception in winter lettuce: 1. Analytical methods and environmental influences. Ann. Bot. 54, 743–757.

Thornley, J. H. M.; Johnson, I. R. (1990). Plant and Crop Modelling. A Mathematical Approach to Plant and Crop Physiology. Clarendon Press, Oxford, 669 pp.

Thornley, J. H. M. and Hurd, R. G. (1974). An analysis of the growth of young tomato plants in water culture at different light integrals and CO2 concentrations: 2. A mathematical model. Annual Botany, 38: 389–400.

6

Precision Farming

Lunagariya Radhika J., M.S. Tondihal and D.R. Kanzaria

6.1 Scenario in precision farming

- India's population is around 1366 million by 2020 which requires food grain production of 350 million tones.
- Considering the current food grain production levels, India needs to raise food grain production levels by not less than 4.00 per cent per year. As the availability of land has decreased, application of fertilizers and pesticides become necessary to increase food production.Indiscriminate use of inputs coupled with improper management practices over a long period has resulted in decline in soil productivity.
- Agriculture production system is outcome of interaction of various agro inputs. Therefore, judicious management of these inputs is necessary for the sustainability of these inputs.
- In our agriculture systems, spatial variability over the landscape is ignored and fertilizers, herbicide, pesticides, fungicides *etc.* are applied at uniform rate over the whole field.
- It has been recognized that crops and soils are not uniform within a given field. The management of in-field variability in soil fertility and crop conditions for improving crop production and minimizing the environmental impact is the crux of precision farming.
- **Definitions**
    - It is defined as the application of technologies and principles to manage spatial and temporal variability associated with all aspects of agricultural production (Pierce and Nowak, 1999).
    - It is an integrated information and production-based farming system that is designed to increase long term, site specific and whole farm production efficiency, productivity and profitability while minimizing unintended impacts on environment.

- Precision agriculture is broadly defined as monitoring and control applied to agriculture, include site specific application on inputs, timing of operations and monitoring of crops and employees.
- It is a new method of farming that tailors inputs of fertilizers, pesticides *etc.* to match the variation in the growing conditions within a field. The practice is known as Site Specific Management.
- It is a concept of using the new technologies and collected field information, doing the right thing, in the right place, at the right time. Collected information may be used more precisely to evaluate optimum sowing density, estimate fertilizers and other input needs.
- It helps in avoiding unwanted practices to a crop, regardless to local soil/climate conditions, *i.e.* it reduces labour, water inputs such as fertilizers and pesticides *etc.* and assures quality produce.
- Hence it is a comprehensive system designed to optimize production by using a key element of information, technology, and management, so as to increase production efficiency, improve product quality, improve the efficiency of crop chemical use, conserve energy and protect environment (Shibusawa, 2002).
- Especially in the case of small farmers in developing countries, precision farming holds the promise of substantial yield improvement with minimal external input use (Fountas *et al.,* 2004).

6.1.1 Traditional farming vs Precision farming

Sr. No.	Precision Farming	Traditional Farming
1.	Farm field is broken into "management zones"	Whole field approach where field is treated as a homogeneous area
2.	Management decisions are based on requirements of each zone.	Decisions are based on field averages.
3.	PF tools (*e.g.* GPS/ GIS) are used to control zone.	Inputs are applied uniformly across a field.

6.1.2 Aim of precision agriculture

- Precision agriculture aims to optimize field level management with regard to:
- Crop science by matching crop practices more closely to crop needs (*e.g.*, fertilizer inputs).
- Environmental protection: by reducing environmental risks and footprints of farming (*e.g.*, limiting leaching of nitrogen).
- Economics: by boosting efficient use of inputs.

6.1.3 Objectives

- Site-specific application of inputs required to grow crops = improve crop productivity.
- Increase in the efficiency of agricultural resources utilization by crops = lower environmental impact.
- Automate and long farm operation data analysis: efficiency and convenience.

6.1.4 Tools and equipment

- Precision farming is a combination of application of different technologies. All these combinations are mutually inter related. The same are discussed below:
    1. Remote Sensing (RS)
    2. Geographical Information System (GIS)
    3. Global Positioning System (GPS)
    4. Yield Monitoring (YM)
    5. Variable Rate Technology (VRT)

6.1.5 Steps involved in precision farming

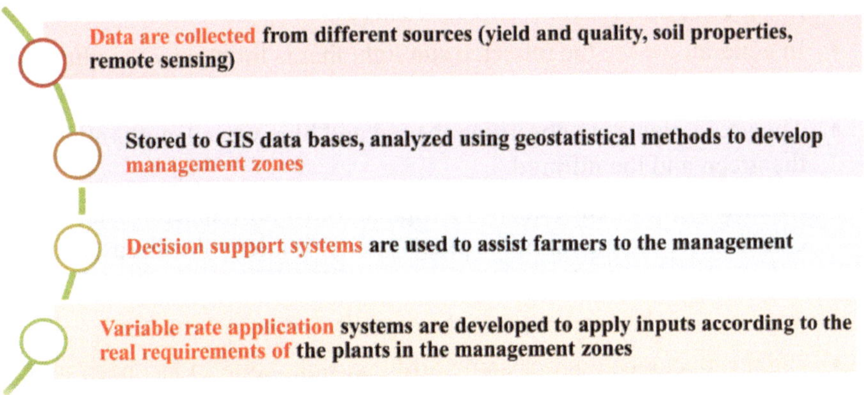

6.1.6 Precision farming development centers in India

- The Precision Farming Development Centers (PFDC) plays a leading role in the development of regionally differentiated precision farming technology validation and dissemination.
- Some important PFDCs in India
    - IARI, New Delhi
    - UAS, Bangalore

- IIT, Kharagpur
- NAU, Navsari
- TNAU, Coimbatore
- CISH, Lucknow

6.2 Remote Sensing

- Remote sensing is a tool for collection, processing and analysis of data to exact information from earth surface without coming in to physical contact with it.
- Satellites used in Remote sensing- RRS-IA, RRS-IIB, RRS-IIIC for agriculture purpose. Now a days, resource satellite (IRS-P6) is used.
- It is a key to detect changes and locating ground positions. RS monitors the spatial variability over the time.
- Remotely sensed data, obtained either by aircraft or satellite, containing electromagnetic emittance data of crop can provide information useful for soil condition, plant growth, weed infestation *etc.*
- This type of information is cost effective and can be very useful for site specific crop management programs.
- It is a useful technology for precision agriculture as can give data for parameters of the field relatively easily.
- In general, we see the reflected sun light that is formed by the ultraviolet wave lengths, the visible light (Red, Green and Blue) and the infrared.
- The green plants are absorbing the red and blue wave lengths and reflect the green and the infrared.
- Measuring the reflected wavelengths with multi spectral cameras we can measure the vigour of the plants or any problem like disease, nutrient deficiency or water logging *etc.*
- We can correlate soil color to the organic matter, moisture *etc.*
- Light reflectance (sun or some artificial light source) has been used in precision agriculture in the form of vegetation indices. The most used of them is the Normalised Vegetation Index (NDVI).
- Several other indices can be calculated and used offering good agreement with certain characteristics of the crop. NDVI has been correlated to crop yield and quality.
- The measurements of plant reflectance can be carried out by satellites, airplanes or ground instruments.
- Derived NDVI data from these maps converted to leaf area index (LAI)

map, which showed potentiality to provide decision support for irrigation and canopy management.

- **The remote sensing processes**

Remote sensing process as applied to disease monitoring processes. The sun (A) Electromagnetic energy (B) To plants (C) A portion of the electromagnetic energy is transmitted through the leaves. The sensor on the satellite detects the reflected energy (D) The data is then transmitted to the ground station (E) The data is analyzed (F) And displayed on field maps.

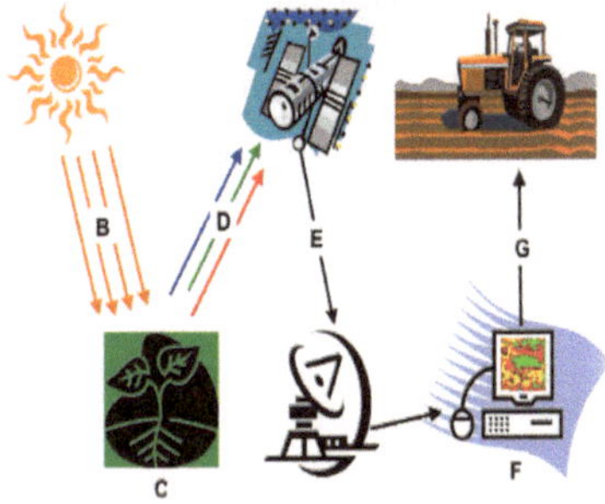

6.3 Geographical Information System (GIS)

- **GIS contributes significantly to precision farming**
  - Allowing presentation of special data in the form of a map
  - Used to manage and analyze spatial data relating to crop productivity and agronomic factors.
  - It contains base maps like topography, soil type, nutrient level, soil moisture, pH *etc.*
  - Data on crop rotation, tillage, nutrient and pesticide application, yields *etc.*
  - A farming GIS database can provide information on filed topography, soil types, surface drainage, subsurface drainage, soil testing, irrigation, chemical application rates and crop yield. Once analysed, this information is used to understand the relationships between the various elements affecting a crop on a specific site (Trimble, 2005).
- **Introduction**
  - GIS is powerful set of tools for collecting, storing, and retrieving the data at will, transforming and displaying the spatial data for particular purpose.
  - The ability of GIS to analyze and visualize agricultural environments and work flows has proved to be very beneficial to those involved in the farming industry.
  - Spatial data are commonly in the form of layers that may depict topography or environmental elements.

- GIS can be used to produce images, not just maps, but drawings, animations, and other cartographic products.
- From mobile GIS in the field to the scientific analysis of production data at the farm manager's office, GIS is playing an increasing role in agriculture production throughout the world.

6.4 Global positioning system (GPS)

- GPS makes use of identify the location of farm equipment within a meter of an actual site in the field.
- Location of soil samples and the laboratory results can be compared to a soil map
- Tillage adjustments can be made as one finds various conditions across the field and
- One can monitor and record yield data as one goes across the field.
- The GPS technology provides accurate positioning system necessary for field implementation of Variable Rate Technology (VRT).
- GPS is a navigation system based on a network of satellites that helps users to record positional information (latitude, longitude and elevation) with an accuracy of between 100 and 0.01 m (Lang *et al.,* 1992).
- The system allows farmers to reliably identify field locations so that inputs (seeds, fertilizers, pesticides, herbicides and irrigation water) can be applied to an individual field, based on performance criteria and previous input applications (Batte *et al.,* 1999).
- **Introduction**
    - GPS is a navigation system based on a network of satellites that helps users to record positional information (latitude, longitude and elevation) with an accuracy of between 100 and 0.01 m.
    - The system was originally designed and maintained by the US Department of Defense (DoD).Location information us collected by GPS receivers for mapping field boundaries, roads, irrigation systems and problem areas such as weeds or disease. GPS allows farmers to accurately navigate to specific locations in field year after year, to collect soil samples or monitor crop conditions.There is an automatic controlling system, with light or sound guiding panel, receiver and antenna.
    - The system allows farmers to reliably identify field locations so that inputs (seeds, fertilizers, pesticides, herbicides and irrigation water) can be applied to an individual field, based on performance criteria and previous input applications.

- Similarly, European Union and European Space Agency (ESA) took initiative to launch their own global navigation satellite system, GALILEO.
- A study was investigated the feasibility of using Real Tie Kinematics (RTK) GPS to map the locations to tomato transplants in the field as they are planted using a transplanter retrofitted with an RTK GPS receiver and real time controller (Sudduth *et al.*, 1997).

6.4.1 Variable Rate Technology (VRT)

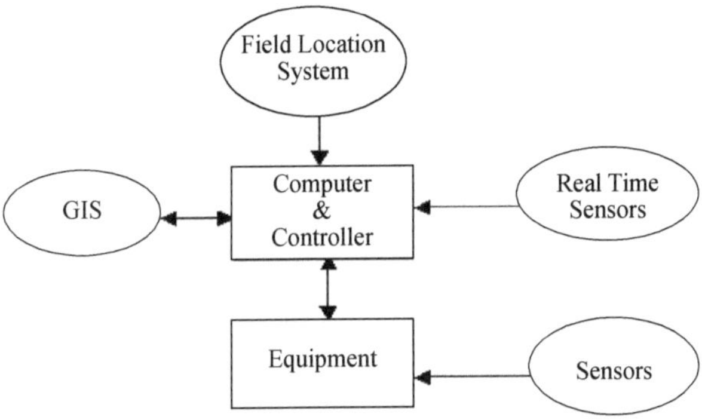

- **Components of VRT systems**
  - VRT is used to adjust the agricultural inputs according to the site-specific requirements in each part of the field.
  - If machines are used, this requires variable-rate machinery.
  - On small farms, inputs can be applied manually.
  - The application map is loaded into a computer mounted on a variable-rate fertilizer spreader. The computer uses the application map and a GPS receiver to direct a product-delivery controller that changes the amount and/or kind of fertilizer product, according to the application map (Berntsen *et al.,* 2006 and Ferguson *et al.,* 2007).
- **Introduction**
  - VR means that the appropriate rates of inputs will be applied leading either to reduced inputs, costs and environmental effects or improved yields and quality.
  - Two methods are used to apply VR.
  - The first called map based, is based on historical data (previous or present year). Process control technologies allow information drawn from the GIS (prescription maps) to adjust fertilizer application, seeding rates, and pesticide selection and application rate, thus providing for the proper management of the inputs.

- The second, named sensor based, uses sensors that can adjust the applications rates on the go.
- The sensors detect some characteristics of the crop or soil and adjust the application equipment.
- VRA can be applied to all inputs.
- Both systems have advantages and disadvantages.
- The on-the-go sensors are more acceptable by the farmers. Probably using a mixture of both will offer most advantages in the future.

6.5 Yield monitoring

- Yield monitor iscropping yield measuring devices installed in harvesting equipment.
- The yield data from the monitor is recorded and stored at regular interval along with positional data revised from the GIS unit.
- GIS software takes the yield data and produce yield map.
- In all yield monitors, GPS receivers are used to record the location of yield data and create yield maps. Other yield monitoring systems include devices used in forage crops to keep track of weight, moisture, and other information on a per-bale basis (Davis *et al.,* 2005).
- **What does a yield map tell us?**
  - "Report card" for field
  - Quantifies variability within afield
  - Directs sampling to explain yield differences
- **Introduction**
  - Yield is ultimate indicator of variation of different agronomic parameters in different parts within the field.
  - So, mapping of yield and interpretation and correlation of that map with the spatial and temporal variability of different agronomic parameters helps in development of next season's crop management strategy.
  - Present yield monitors measure the volume or mass flow rate to generate time periodic record of quantity of harvested crop for that period.
  - Time periodic yield data is then synchronized with location address obtained from onboard GPS system to create most common color-coded thematic map.
  - Yield mapping can be carried out easily in mechanized crops.

6.6 DRIS and SSNM for precision farming in horticulture

- Diagnosis and Recommendation Integrated System (DRIS) represent a holistic approach to the mineral nutrition of crop and has an impact on the integrated set of norms representing calibration of plant tissues, soil composition, environmental parameters and farming practices as the functions of the yield of a crop.
- Once such norms are developed, it is possible to make a diagnosis of the conditions of the crop thereby isolating the factors, which are likely to be responsible for limiting the growth and production.
- The most important advantages of DRIS approach are its ability to make a diagnosis at any stage of crop development and to list the nutrient in the order of importance, which are responsible for limiting the yield.
- Diagnosis and recommendation integrated system has been developed to fulfil the diagnosis and predictive use of the leaf analysis.
- Optimization these factors create conditions, which are likely to the chances of obtaining higher yield and quality.
- DRIS employs survey technique, where a large number of randomly distributed sites throughout the area are selected. At each site, the samples of the leaf and soil are taken for the analysis and the details of the applied manures and fertilizer are recorded.
- Site-specific management differs from the traditional practice of whole field management.
- In whole field management, average conditions for a field or a farm are determined and management practices applied accordingly.

6.7 Advantages of precision farming

- Reduced cost of cultivation.
- Environment pollution.
- Application of nutrients.
- Chemical doses through variable rate application technology.
- Application of irrigation water, thus reduce the leaching of nutrients along with deep percolation. runoff, erosion and sedimentation of water bodies.
- Minimum inputs to get maximum output.
- Increase in production efficiency of inputs

6.8 Constraint of precision farming in India

- Small land holdings
- Poor socio-economic status of farmers
- Knowledge and technology gap
- Heterogeneity in cropping system
- Lack of local technical expertise
- Lack of data availability market performance
- Unavailability of data in terms of quality and costs

6.9 Application in fruit and vegetable crops

- Growing media
- Water management through,
  - Drip irrigation
  - Fertigation
- Mulching
- High density planting
- INM
- Plant growth regulators
- Varietal performance
- Intercropping
- Biotechnology/Micro propagation/Tissue culture
- Economics
- In fruit and vegetable farming the recent rapid adoption of machine vision methods allows growers to grade products and to monitor food quality and safety, with automation systems recording parameters related to product quality. These include colour, size, shape, external defects, sugar content, acidity, and other internal qualities (Njoroge *et al.,* 2002). Additionally, tracking of field operations such as chemicals sprayed and use of fertilizers can be possible to provide complete fruit and vegetable processing methods.
- This information can be disclosed to consumers for risk management and for food traceability as well as to producers for precision agriculture to get higher quality and larger yields with optimized inputs. In recent years several new approaches were developed that take into account the actual size of the tree, the condition of the crop, but also the environmental conditions (Doruchowski *et al.,* 2009).

- Grape quality and yield maps are of great importance during harvest to avoid mixing grapes of different potential wine qualities. The parcels with greatest opportunities for PV are those which reveal a high degree of yield variation. A high degree of variation will mean higher VRA of inputs and, therefore, greater economic and environmental benefit in comparison with uniform management.
- The development and adoption of PA technologies and methodologies in viticulture termed Precision Viticulture (PV) is more recent than in arable land. However, driven by the high value of the crop and the importance of quality, several research projects already exist in wine production areas of the world (Ojeda *et al.,* 2005 and Ferreiro-Arman *et al.,* 2006).

References

Batte, M. T. and Van Buren, F. N. (1999). Precision farming – Factor influencingproductivity. In Northern Ohio Crops Day meeting, Wood County, Ohio, 21 Jan.

Berntsen, J.; Thomsen, A.; Schelde, K.; Hansen, O. M.; Knudsen, L.; Broge, N.; Hougaard, H. and Horfarter, R. (2006). Algorithms for sensor-based redistribution of nitrogen fertilizer in winter wheat. Precision Agriculture,7: 65-83.

Davis, G.; Massey, R. and Massey, R. (2005). Precision agriculture: An introduction. www. muextension.missouri.edu/explore/envqual/wq0450.htm.

Doruchowski, G.; Balsari, P. and Zande, J. C. (2009). Precise spray application in fruit growing according to crop health status, target characteristics and environmental circumstances; Proc. of 8th Fruit, Nut and Vegetable Production Engineering Symposium, Concepcion-Chile. pp. 494-502.

Pierce, F.J. and Novak, P. (1999). Aspects of Precision Agriculture. Advances in Agronomy, 67: 1-85.

Ferguson, R.; Dobermann, A. and Schepers, J. (2007). Precision agriculture: site-specific nitrogen management for irrigated corn. University of Nebraska Lincoln Extension. Bulletin.1-7.

Ferreiro-Arman, M.; Da Costa, J. P. and Homayouni, S. (2006). Hyperspectral image analysis for precision viticulture, In Image Analysis and Recognition, Springer Berlin Heidelberg. pp. 730-741.

Fountas, S.; Ess, D.; Sorensen, C. G.; Hawkins, S.; Pedersen, H. H.; Blackmore, S. and Deboer, L. J. (2004). Farmer experience with Precision Agriculture in Denmark and US Eastern Corn Belt. Precision Agriculture.

Lang, L. (1992). GPS, GIS, remote sensing: An overview. Earth Observation Magazine. pp. 23-26.

Njoroge, J. B.; Ninomiya, K. and Kondo, N. (2002). Automated fruit grading system using image processing, In Proceedings of the 41thSICE Annual Conference. pp. 1346-1351.

Ojeda, H.; Carrillo, N. and Deis, L. (2005). Precision viticulture and water status II: Quantitative and qualitative performance of different within field zones, defined from water potential mapping, in XIV International GESCO Viticulture Congress, Geisenheim, Germany, pp. 741-748.

Shibusawa, S. (2002). Precision farming approaches to small farm agriculture. Agro-Chemicals Report,2(4):13-20.

Sudduth, K. A.; Hummel, J. W. and Birrel, S. J. (1997). Sensors for site specific management. Am. Soc. Agron. Madison, WI, 183-210.

Trimble, (2005). Precision agriculture. Available at: www.trimble.com.

7

Advances in Crop Regulation Techniques Infruit Crops

V.M. Parmar, Mithapara Kinnari D. and Makavana Payal C.

Introduction

- India is the second largest producer of fruits in the world after China
- India is rich in fruit diversity starting from tropical, subtropical to temperate region.
- Some of the fruits like Guava, pomegranate, lemon, mandarin *etc.*, if left without any treatment, give several light harvests of the variable quantities and qualities from the various flowering flushes throughout the year.
- Some of the fruit crops bloom throughout the year without any resting period and produces two or three crops (bahar) in a year but yield and quality is not so good in all crop harvest. So, it is very essential to understand the flowering and fruiting behavior of crops and which bahar will give good crop with considering all the factors associated with a particular bahar.

7.1 What is crop regulation?

- Crop regulation means to force the tree for its rest and to produce blossom and fruits during any one of two or three flushes"
- Crops in which flowering is continuous throughout the year; economic yield cannot be achieved in any of the season.
- Therefore, by analyzing market demand flowering can be forced in particular one season, which gives the economic yield.
- Through various techniques like cultural practices and chemical treatments, trees are thrown to rest prior to flowering and then practices are done to trigger flowering.
- Thus, crop regulation is a method to regulate flowering by altering natural tendency of the tree by artificial means.

How crop regulation is useful?

- Crop regulations treatment is important in the determination of crop yield, some events occurring during flower formation and set affect fruitlets development and final fruit size and quality, having an additional effect on returns.
- Understanding of crop regulation will help to optimize harvest time which reduce glut in the market, to obtain higher price of the produce, to avoid spoilage due to excessive production at a particular time, to provide employment throughout the year and to reduce imports and balance the trade.
- Moreover, it will also help in pest and disease control by avoiding exposure of plant susceptible stage with respect to particular pest/disease cycle.
- **Why crop regulation?**
  - Many crops which bloom more than once in a year do not produce good yield and quality of the fruit throughout the year. The yield and quality depend on the bahar.
  - For example, the rainy season crop of guava is poor in quality and crop is affected by many biotic and abiotic stresses as compared to winter season crop. The winter season crops (mrigbahar) which ripen from second fortnight of October to first fortnight of January are superior in quality.

7.2 Principle of crop regulation

- The basic principle of crop regulation is to manipulates the natural flowering of the plant in desired season that contribute to increased fruit yield, quality and profitability.
- This concept is based on the fact that most of the crops' flowers are borne only on new, succulent, vigorously emerging vegetative growths.
- These new growth flushes can be either on new emergences of lateral bud on older stems or extensions of already established terminals of various size and vigour.

7.3 Objectives of crop regulation

- The main objective of crop regulation is to force the tree for rest and produce profuse blossom in desired season
- To take the harvesting early or late than that of normal growing season
- To meet the fruit demand in off-season and to sell the product with higher price

- Production is continued throughout the year and not related to the climatic condition
- To reduce cost of cultivation as continuous blossom would produce light crops throughout year and require a high cost for the monitoring and marketing
- **The selection of bahar is determined by**
  - Availability of the irrigation water
  - Quality of products
  - Occurrence and extend of the damage by the disease and pests
  - Market demands
  - Climate of the area
  - Availability of fruit in the market
  - Comparable yields
- **Types of bahar**

Bahar	Time of flowering	Time of fruiting
Ambebahar	January to February	Rainy season – June to August
Mrigbahar	June to August	Winter season – October to December
Hasta bahar	October to November	Summer season – February to April

7.4 Methods of crop regulation

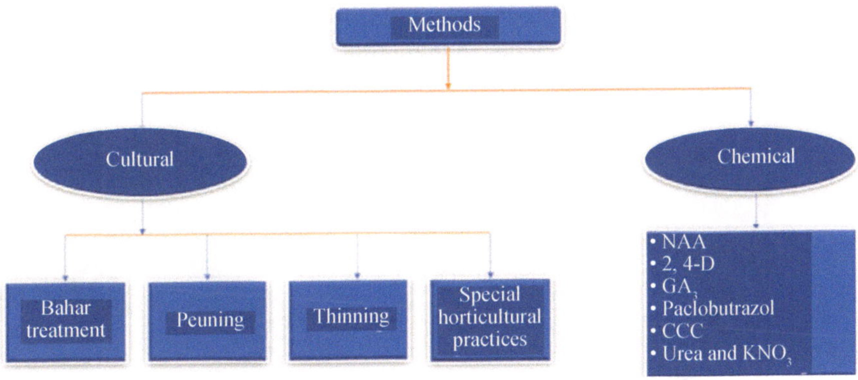

7.4.1 Cultural methods

1. **Withholding irrigation**
   - The principle behind withholding irrigation is to suspending the vegetative growth and increasing reproductive growth by providing rest to the plant, which results in accumulation of food in large quantity for enhancing flowering in the next season.

- In case of mrigbahar in guava, induction of water stress by withholdingirrigation from April - June or until the beginning of monsoon depending upon the prevailing condition
- During rest period, trees can conserve food material in its branches. In the month of June tree is cultivated and manured followed by irrigation.
- After about 25-30 days tree would produce into profuse flowering.
- The fruits will be matured during winter.

2. Pruning

- Pruning of terminal shoots can produce off season crops by delaying flowering time
- In guava pruning of terminal portion up to 20-30 cm length of current season growth of spring flush is effective to avoid the rainy season crop
- It is done in April or May depending upon the location
- The pomegranate requires time to time pruning throughout its life cycle
- Fruits are borne terminally on short spurs, arising from matured shoots, which have capacity to bear fruits for 3 to 4 years
- A little thinning or pruning of old spurs to encourage the new ones is required for regular flowering and fruiting
- In acid lime pruning helps in induction of hasta bahar when its demand is high.

3. Hand thinning of flowers

- Flower thinning with hand is found to be most effective method for regulation of flowering and quality fruit production.

- Hand thinning of flowers/fruits in guava during March-April which reduces the rainy season crop and thereby increasing the winter season crop.
- Pomegranate plant can set too many flowers and fruits but it causes overloading bending or breaking off the branch from plant.
- Hence 'thinning effect' is given to tree by removing some flowers and immature fruits in earlier stages of Bahar.

4. **Bending in guava**
    - Shoot bending is one of the ways to produce better quality fruits in the off-season of guava
    - In case of bending of branch wood tension of branch is increased and phloem formation decreased
    - As a result,photosynthetic product pass slowly from the shoots of bent branch as to the other parts, that increased C:N ratio and induce more flowering and fruit set
    - Bending forced dormant reproductive buds into growth
    - Bending of guava employed at the age of 2nd year

Bending

5. **Girdling**
   - Girdling has been, and is still, worldwide used for centuries in citrus, mainly to increase flowering, fruit set and fruit size
   - Girdling consists of removal of a strip of bark from the trunk or major limbs of a fruit tree, thereby blocking the downward translocation of photosynthates and metabolites through the phloem
   - Girdling helps in accumulation of food above the girdle

   - On the upper part of girdling leaf N content, C:N ratio and carbohydrate were improved. Therefore, flowering and fruit set were increased
6. **Exposure of roots and root pruning**
   - In this method, roots of the plant are exposed to sun by removing up to 7-10 cm soil around 40-60 cm radius of tree trunk.
   - As a result of water stress, leaves show wilting and fall on ground.
   - Before one month of flowering (3 to 4 weeks) of desired bahar, roots are again covered with mixture of soil and FYM and light irrigation is given.
   - Exposing feeder roots and pruning of fibrous ones can also induces the flowering in desired season.

7.4.2 Chemical methods

- Growth regulators and certain chemicals have been found very effective in thinning of flowers and manipulating the cropping season.
- NAA, 2,4-D, Bromouracil and Ethrel in variable doses are effective for induction and regulation of the fruit crop in different agroclimatic condition and found successful in reducing the rainy season crop and increasing the winter crop.
- Foliar application of NAA effectively controls the fruit drop and reduce the tendency of alternate bearing.
- Ethrelsynchronizing flowering and also regulate the type of flowers in pomegranate. It increases the hermaphrodite flowers production and also impart the natural color to the fruits during unfavorable climatic conditions.

- Foliar spraying of Urea in April-May reduces the rainy season crop and increases the winter season crop.
- Gibberellins stimulates cell division and cell elongation in plants.
- It influences flowering in citrus as the root - produced gibberellins play an important role in flowering through inhibiting effect on generative shoot production.
- In the recent years, paclobutrazol a growth retardant has been used with considerable success to induce early flowering for off-season production in several fruit crops.
- Cycocel, which act as an anti-gibberellin compound and arrested the vegetative bud development,nucleic acid synthesis and protein metabolism by specific anti-metabolites, which induce flower formation.
- Potassium, zinc and iron are known for development of fruit, movement of sugars and indirectly photosynthesis process. It regulates the opening and closing of the stomata, which is essential for photosynthesis, water and nutrient transport.

7.5 Crop regulation in guava

- Guava is most important commercial fruit crop grown in sub-tropical region of the Indian subcontinent.
- In guava, two distinct seasons of flowering, spring (March-April) and rains (June-July) occur from which fruits ripen during rainy and winter season respectively.
- In North Indian climate the rainy season crop of guava is poor in quality and nutritive value and is affected by many insect pests and diseases.
- The winter season fruits are superior in quality free from diseases and pests and give higher income. But it is advisable to take only one crop every year.
- This requires management of flowering to obtain the most desirable crop, by the methods like withholding irrigation, pruning, thinning of flowers by chemically or manually.
- In guava, mainly mrigbahar flowering is preferred by grower and fruits are harvested during winter season. This season crop is free from insect pest and diseases, and produces good quality fruits (Lal and Das, 2017).

7.6 Crop regulation in pomegranate

- Pomegranate flowers continuously when watered regularly.
- The plants under such conditions may continue bearing flowers and bear small crop irregularly at different period of the year, which may not be desirable commercially.
- To avoid this, trees are given bahar treatment.
- In this treatment, the irrigation is withheld two months prior to the bahar followed by light earthing up in the basin. This facilitates the shedding of leaves. The trees are then medium pruned 40-45 days after withholding irrigation. The recommended doses of fertilizers are applied immediately after pruning and irrigation is resumed. This leads to profuse flowering and fruiting. The fruits are ready for harvest 4-5 months after flowering.
- In tropical condition, there are three flowering seasons, *viz*., January-February (ambebahar) June-July (mrigbahar) and September-October (hasta bahar).
- The choice of flowering/fruiting is regulated taking into consideration the availability of irrigation water, market demand and pest/disease incidence in a given locality.
- The fruits of ambebahar are ready for harvest in the month of June to September. As the fruit development takes place during dry months, they develop an attractive color and quality thus suitable for exports. Similarly due to dry weather, the incidences of pest and disease attack are limited. However, ambebahar can be taken only areas having assured irrigation facilities.
- The mrigbahar crop is harvested in the month of December to February. Usually, this bahar is favoured as the flowering and fruiting period coincides with rainy season or immediately after rains, and the crop is taken without much irrigation. As the fruits develop during the rainy season and mature during winter, the color and sweetness of the fruit is affected.
- The fruits from hast bahar are harvested during the month of March to April.
- They have very attractive rind with dark-colored arils.
- Since the availability of the fruits during this season is limited, they fetch high value. Optimum water stress cannot be developed during this period as withholding of irrigation coincides with the rainy season. This leads to poor flowering and thus affects the yield.

7.7 Crop regulation in acid lime

- The lemon, lime and citron are considered as continuous bloomers, particularly under tropical climate producing some flowers throughout the year, though the spring bloom is the heaviest. The acid limes bloom throughout the year but the main blooming period is February - March, with lean period from July to August (Athani *et al.,* 1998).
- Acid lime in tropical and sub-tropical conditions tends to give continuous flushes of growth, both vegetative as well as reproductive throughout the year.
- Acid lime trees flower thrice in a year in the months of January-February, June- July and September-October known as Ambe, Mrig and Hast bahar, respectively.
- The fruits of the Ambe, Mrig and Hast bahar flowering becomes available in the month of June-July, November-December and April-May months, respectively (Barbera and Garimi, 1988).
- The fruits of Hast bahar flowering become available in the months of April-May when there is heavy demand and is sold at premium price. But Hast bahar (summer cropping) bear only 17 % flowering and fruiting is achieved in the uncontrolled condition because of the monsoon rains preceding flower initiation. Therefore, acid lime is forced to produce hast bahar crop (Ghawade*et al.,* 2002).

7.8 Effect of crop regulation treatments on physiology

7.8.1 Stress physiology

- Stress is any external factor that results in less than optimum growth rates i.e. any factor that interrupts, restricts or accelerates normal metabolic processes of plant or its part.
- Horticultural crops worth many billions dollar are seriously damaged or destroyed annually by pest, disease, drought, floods, wind storms and freeze.
- However, great recognition has been given to the manifold beneficial ways in which natural or induced stress has been and is being utilized.
- Hardening-off of young plants through drought stress and other controlled exposure to adequate illumination to produce sturdy growth, light wounding or girdling to induce flowering or fruit ripening, and controlling temperature and other conditions for product storage are examples of controlled stress for useful ends.

- Every plant and its parts are subjected to stress of some sort to some degree during its lifetime, yet there are few sharply defined limits below which a given stress can be beneficial or useful and above which it is detrimental or lethal.
- The earliest applications of mechanical stress date back to Roman days when Virgil (First century BCE) and Columella (First century) described methods of girdling, ringing, scoring and notching to dwarf plants, to induce flower bud formation and to promote fruiting, with consequent increased fruit size, higher sugars and maturity a few days to a week earlier.
- Girdling is standard practice for grapes in many parts of the world and is widely used in commercial tree fruit culture, mainly for overly vigorous or late fruiters.
- Notching just above a bud will force it into vegetative growth and notching just below a bud will promote fruit bud initiation.
- Girdling or ringing in which a strip of bark is removed, scoring, which is a cut into the wood without removing the bark and various forms of ligature have been carried out also in tropical and subtropical fruit crops, for example on mango, citrus in Israel and Florida.
- Unfortunately, repeated (annual) girdling of the same tree results in progressively less response. Beneficial effects from this intentional wounding have been explained in terms of preferential accumulation of photosynthates.
- Moreover, it will also change the amount and redistribution of growth regulators, particularly with respect to the accumulation of acidic gibberellins in the tissues above the girdle and these effects also mediated by ethylene.
- Bending and tying or weighting down branch ends with a brick, planting trees at a 45° angle, bending and twisting branches into a loose knot have all been used in stress induction of growth of otherwise dormant lateral buds and to promote flower-bud formation and fruiting.

7.8.2 Water stress physiology

- Moderate plant water deficits enhance flowering in many important horticultural crops such as litchi and citrus.
- Flowering intensity is found to negatively correlate with relative water content.

- The water stress is created through deficit irrigation methods and so far, there are two methods of deficit irrigation:
- One is regulated deficit irrigation (RDI) where a certain percentage of evapotranspiration (ET) is replaced by irrigation applied over the entire root zone.
- The other, is partial root zone drying (PRD) where at each irrigation time only part of the root system is watered while the other side is left to dry, to a predetermined level, before being irrigated next.
- The root system senses soil drying and produces chemical signals that are transmitted to the shoots to close the stomata (decreasing water loss) and limit vegetative growth, thus improving WUE.
- There is evidence suggesting that the plant hormone abscisic acid (ABA) is a major component involved in the control of stomatal conductance as the soil dries.
- To avoid die off of the roots and in order to induce a new hormonal stimulus, the irrigated and the dry side are normally changed in a 10–14 days cycle.
- Regulated Deficit Irrigation has been used successfully to maintain yield and fruit quality in many fruit species.
- It has been mentioned above that water stress is a strong inducing factor of flowering but aside from this stimulating effect many other physiological parameters are negatively affected by water stress.
- Characteristic leaf injuries and abscission are typical consequences of water deficit conditions. After long exposure to water deficit, gas exchange parameters, CO_2 assimilation, water potential and stomatal conductance are progressively affected.

7.8.3 Girdling physiology

- The horticulture science is geared towards production of high-value crop, integrating quantity and fruit quality.
- This is achieved by various techniques, including breeding, nutrition, pest and disease control and bioregulators as well as direct manipulation of the plant itself.
- Direct plant manipulations leading to the desired yield consist of two kinds of horticultural techniques:
    - removal of certain tree organ (*e.g.* pruning, fruit thinning) and
    - interference with translocation between major tree organs (*e.g.* girdling, ringing, branch bending which modifies auxin distribution, *etc.*)

- Farmers have practiced the use of girdling and related techniques (cincturing, hacking or frilling, notching, nicking, ringing, scoring, strangulation, stripping and wiring) in horticulture for thousands of years in order to increase crop production.
- Girdling is basically an intervention in the phloem transport between canopy and roots, in an attempt to manipulate the distribution of photosynthates, mineral nutrients and plant hormones.
- Two main factors limit the extensive use of girdling:
  1. difficulties related to determining the optimal timing and environmental conditions for each species and location; and
  2. uncertainties concerning the effect of girdling on trees and fear of causing severe or even lethal damage by single or repeated treatments.
- Girdling interfered with the movement of sucrose into roots and at the same time reduced the export of cytokinins from the roots to the shoots. Thus, girdling seems to remain an efficient tool for growth regulation in both herbaceous and perennial plants (Wick and Davenport, 1987).

7.8.4 Pruningphysiology

- There are several reasons for pruning of the plant in the orchard.Pruned trees restore the balance between the root system and the above ground part.Young trees are usually pruned in order to obtain a framework that is strong and accessible for other cultural practices.
- Fully grown fruit trees are pruned to maintain the canopy height, spread and density required for easy spraying, fruit thinning and harvest.
- The very first experiments on pruning showed that fruit trees left entirely without pruning or pruned lightly produce many more flower buds in their early years than heavily pruned trees.It is evident from many studies that moderate or slow growth of shoots is indispensable to initiation of flower bud formation.
- Hormones supplied by the roots are unbalanced when the top parts are removed and the hormones produced by the green parts of the tree cannot be utilized when the number of meristems is reduced.It may be due to that auxins produced in actively growing shoot tips which inhibit flower bud formation.Restriction of shoot growth by such treatments as low nitrogen fertilization, application of growth retardants or shoot bending may promote and increase flower bud formation.
- In addition to growth hormone level, which is strongly related to the intensity of growth, the supply of assimilates seems also to play an important role in flower bud formation. Dormant pruning may decrease

the pool of available assimilates as a results of reduced foliage area or increased demand of activated meristem.

- Restricting the assimilate supply by shading or defoliation decreases or even disrupts flower bud formation.On the other hand, increasing the assimilate supply in shoots by ringing or shoot bending increase flower bud formation.

References

Athani, S.I.; Hulamani, N.C. and Patil, M.P. (1998). Flowering and fruit set behaviour of kagzi lime strains. Adv. Agric. Res., 9:151-53.

Barbera, G. and Garimi, F. (1988). Sixth Intern. Citrus Congress Middle East Telavive Israel, 8 - 11 March. 88: 2.

Ghawade, S.M.; Panchbhai, D.M.; Jadhao, B.J.; Katole, S.R. and Dadmal, S.M. (2002). Flowering behaviour of kagzi lime (Citrus aurantifolia Swingle) under sub-humid tropical climate. Agric. Sci. Digest., 22: 252-54.

Lal, N. and Das, R.P. (2017). Effect of Plant Growth Regulators on Yield and Quality of Guava (*Psidium guajava* L.) cv.Allahabad Safeda. Int.J.Curr.Microbiol.App. Sci., 6(5): 857-863.

Wick, S.M. and Davenport, T.L. (1987).Proc. of plant growth regulator. Soc. of Amer., 487-488.

8

Influence of Stress Factors and Strategies to Overcome Stress Effect

M.S. Tondihal, H.N. Patel and F.P. Karagatiya

8.1 What is stress?

- **Physics:**Mechanical force per unit area applied to an object.
- **Biology:**Stress is any adverse environmental factor or condition that affects normal metabolic or physiological processes.

  It may be :

  - Biotic stress (living)
  - Abiotic stress (non-living)

8.2 Abiotic stress

- Abiotic stress is defined as:"The negative impact of non-living factors on the living organisms in a specific environment"
- Only 10% of the world arable land may be categorized as free from stress.

8.2.1 Type of abiotic stress

8.2.1.1 Temperature stress

- Temperature is one of the most important components of climate. It plays vital role in the production of horticultural crops.
- The different activities of plant like growth and development, respiration, photosynthesis, transpiration, uptake of nutrients and water and reproduction (Such as pollen viability, blossom fertilization fruit set *etc.*), carbohydrate and growth regulators balance, rate of maturation and senescence, and quality, yield and shelf life of the edible products.
- The above function of the plant should be well when the temperature at the optimum range.During high temperature plant does not perform proper functions of growth, where in low temperature physiological activities of the plant are stopped.

- According to different temperature range in the tropics, the specific trees are grown in different location *e.g.* apple, pear, peach, almond are successfully grown in the regions of low temperature known as temperate fruits.
- In warm winter areas, due to insufficient chilling temperature fruit trees fail to complete their physiological rest period or meeting their chilling requirement.
- As a consequence, buds remain dormant, and leave and blossoms do not appear on the trees in the following spring.For this reason, temperate fruit like apple, apricot, pear and plums are not considered suitable for tropical or subtropical regions.
- For tropical and sub-tropical fruits, the minimum temperature must be within the limit of tolerance of the fruit species.
- The high temperatures affect banana growth and production. The leaf production and relative leaf area growth was affected beyond 33.5°C (Turner and Lahav, 1983). Hightemperature (31-32°C) shortening the bunch development period by increases the rate of plant maturity in Banana (Turner *et al.*, 2007).
- Banana plant can relatively persist under prolonged water stress, but the combined effects of deficit soil moisture along with prolonged prevalence of temperatures beyond 35°C can reduce banana production (Thornton and Cramer, 2012).
- In guava poor fruit set occurs in north India during summer month where climate is characterized by high temperature and low humidity (Chadha and Pandey, 1986).
- In Japanese pear high temperature during early fruit development of water core disorder (Sakuma *et al.*, 1995).
- In custard apple no fruit set due to air temperature more than 40°C (Rao *et al.*, 2016).
- Higher temperature (28/18°C max/min.) development of translucency disorder in pineapple fruit (Paull and Reyes, 1996).
- The fruit grown in tropical and sub-tropical climate is known as tropical fruit and sub-tropical fruits. The activities of the plant are affected by very high or very low temperature. The temperature range for plant is
  - Minimum: 4.5° to 6.5°C
  - Optimum: 24° to 27°C
  - Maximum: 29.5° to 45.4°C

- **Effect of high temperature on plants**

  Mechanisms:

  - Excessive membrane fluidity.
  - Disruption of protein function.
  - Metabolic imbalances.
  - Disruption of normal mRNA precursors.
  - Disruption of normal protein synthesis.

- **Affect crop at high temperature**

I. Vegetative growth

- Increase in vegetative growth and after that ceases the growth.
- High temperature exposure (43°C) in almond bud failure like symptoms appear in which there is complete separation of distorted and compressed cells from surrounding tissues takes place by a periderm like layer.

II. Flowering

- Abnormal pattern of bud break and development in temperature fruit trees.
- In apricot, the warm temperature (15.9°C) 3-5°C higher than the normal results in underdeveloped pistils.

III. Fruit set and Yield

- The higher temperature during fruit set stage also affects the fruit retention.
- On trifoliate orange rootstocks, when exposed to 30-35°C for 48h in a controlled environment increased the fruit drop. It was found that following exposure, IAA concentration increased in the leaves but decreased in the fruitlets. Conversely, there was little effect on ABA in leaves but ABA concentration and ethylene produced at higher temperature responsible for fruit drop.

IV. Metabolic processes

- High temperature also has a direct effect on the processes like respiration and photosynthesis.

- **Effect of high temperature in fruit crops**

1. Spongy tissue in mango

- It is specific in Alphonso mango. This disorder reduces exports of this variety.

    - The tissue become spongy due to inactivation of ripening enzyme induced by high temperature, convective heat and post-harvest exposure to sunlight.

- **Management**
    - Use of sod culture and mulching can reduce this disorder.
    - Varieties like Ratna and Arka Puneet which are hybrids of Alphonso do not suffer from this malady.

2. **Fruit drop in citrus**
    - Excessive pruning can stress the citrus tree and remove too much of its foliage, forcing fruit drop.
    - A severe pest infestation can also stress the tree enough to cause fruit drop.
    - Sudden changes in temperature, particularly when high temperatures occur at or shortly after fruit set, can cause fruit drop.

- **Management**
    - Spray of some hormones on blooms ensures fruit setting.
    - Application of 2, 4-D-10 ppm combined with aureofungin at the first week of fruit drop.

3. **Sunburn in apple**
    - Sunburn in apples is caused by excessive solar radiation and high temperatures.
    - Burning of leaves and fruits.Transpiration losses are also much higher at that time.
    - The bark of the affected trunks and branches may crack.

- **Management**
    - Use more heat tolerant fruit varieties.For example: Pink Lady and Sundowner are more tolerant than Granny Smith and Gala.
    - Shade netting is used to control sun damage in apple orchards.
    - Over-tree sprinkler, cooling systems are designed to reduce sunburn by delivering misted or sprinkled water over the orchard canopy to cool fruit during the hottest part of the day.
    - An additional benefit of over-tree sprinkler cooling is that it can increase color development in red-blush apples.

4. **Sun scald in guava**
    - Guavas exposed to direct sun light may be scalded.

- **Management**
    - In some countries, paper bags are used to cover guava fruits and protect them from solar radiation and insect infestation while on the tree.

5. **Heat injury in papaya**
    - Exposure of papayas to temperatures above 30°C for longer than 10 days those needed for decay result in heat injury (uneven ripening, blotchy ripening, poor color, abnormal softening, surface pitting, and accelerated decay).

- **Management**
    - Quick cooling to 13°C (55°F) after heat treatments minimizes heat injury.
- **Affect crop at low temperature:**
    - Cold-related stress can be defined in terms of either chilling (between 0°C and 12°C) or freezing or below 0°C without snow cover.
- **Stress by cold temperature**
    - Low temperature stress
    - T> 0 °C: Chilling
    - T< 0 °C: Frost
    - Membrane viscosity
    - Retarded metabolism
    - Delayed energy dessipation, leading to radical formation and oxidative stress
- **Stress by freeze dehydration**
    - Protoplast volume shrinkage upon extracellular ice formation
    - Negative turgor
    - Concentration of cellular solutes
    - Change of membrane potentials
    - **Effect of low temperature in fruit crops**

1. **Clustering in Mango**
    - The clustering (Jhumka) at the tip of the panicle and appears as of bunch tip.
    - Clustering develops due to adverse climate mainly low temperature in February-March.

- Most of the fruits are aborted with shriveled embryos and do not develop further, signifying the role of normal embryo growth in the development of fruits.

- **Management**
  - Introduction of beehives in the orchards during flowering season for increasing the number of pollinators.
  - Spraying of NAA (300 ppm) during October-November is recommended

2. **Choke throat in banana**
   - It is due to low temperature affecting active growth of the plant. Leaves become yellow and in severe cases, the tissue gets killed.
   - The distal part of the inflorescence comes out and the basal part gets stuck up at the throat. Hence, it is called *Choke throat.*

- **Management**
  - Provision of shelter belts using Casuarina or Eucalyptus to prevent the effect of cold wind blowing into the orchard and planting low temperature tolerant varieties like Kullan to check the disorder.

8.2.1.2 Water stress

- **Two main conditions develop water stress in plants**
  - **Waterlogging** refers to the saturation of soil with water. Soil may be regarded as waterlogged when the water table of the groundwater is too high to conveniently permit an anticipated activity, like agriculture. In agriculture, various crops need air (specifically, oxygen) to a greater or lesser depth in the soil.
  - **Drought** can be defined as the absence of rainfall or irrigation for a period of time sufficient to deplete soil moisture and injure plants.
- **Plant responds to water stress by closing their stomata**
  1. Drought
  2. Increased ABA
  3. Increased intercellular cytosolic calcium
  4. Inhibition of inward K^+ channel going to guard cells
  5. Loss of turgor
  6. Stomatal closure
- **Effect of drought stress on plants**
- **Effect on growth**

  - Reduction of turgor pressure, due to cell size will be smaller.
  - Effect on photosynthesis:
  - Photosynthesis decreases due to disruption of PS II (photo system II), stomatal closure and decrease in electron transport.
  - Decrease in nuclear acid and proteins Protease activity, RNAase activity, RNA hydrolysis and DNA content falls down.
- **Effect on Nitrogen metabolism**
  - Nitrate reductase activity, nitrite reductase activity insensitive.
- **Effect on carbohydrate metabolism**
  - Loss of starch and carbohydrate translocation decrease.
- **Major reasons of drought**
- **Atmospheric factors**
  - High temp
  - High wind
  - Air pollution
- **Soil factors**
  - Low temp
  - Excessive soil salinity
  - Receding water content
- **Root signaling under drought stress**
  - The root:shoot ratio of plants increases, because roots are less sensitive than shoots to growth inhibition by low water potentials.
  - Roots induce a signal cascade to the shoots via xylem, causing physiological changes.
  - ABA, cytokinins, ethylene, malate and other unidentified factors have been implicated in the root–shoot signaling.
  - This drought-induced root-to-leaf signaling through the transpiration stream results in stomatal closure, which is an important adaptation to limited water supply in the field.
- **Effect of drought in fruit crops**

1. **Fruit drop in mango**
   - Climatic conditions like disturbed water relations, lack of nutrition, competition among fruits and pests and diseases, contribute to fruit drop, further, hormones and their activities in the seeds play an important role in deciding the retention of fruits.

- **Management**
  - The extent of fruit drop in mango can be significantly reduced by regular irrigations during fruit development period.
  - Orchards should be protected from desiccating winds by planting wind breaks.
  - Timely and effective control measures against major pests and diseases should be adopted.
  - Growth regulators like NAA and 2, 4-D in varying concentration depending upon the variety and time of application can be sprayed. The optimum concentration lies between 10 and 15 ppm during the month of April-May.
  - Various cultural management practices, optimum application of irrigation at appropriate time and mulching during fruit development stage as well as exogenous application of plant growth regulators and nutrient at proper dose in appropriate time (Lunagariya *et al.* 2023).

2. **Fruit splitting in fruit crops**
   - This is mainly a physiological disorder and is largely attributed to high atmospheric humidity following heavy rains or heavy irrigation during hot weather.

- **Management**
  - The disease can be minimized by frequent and light irritations during the dry and hot periods and early picking of fruits soon after maturity.

8.2.1.3 Salinity stress

- **Salt stressoccur due to**
  - Poor soil drainage
  - Improper irrigation method
  - Poor water quality
  - In sufficient water supply and rise in the ground water table.
  - Saline soils normally occurs in Punjab, Haryana, U.P., Gujarat, Rajasthan, West Bengal, Maharashtra, AP., Karnataka and Tamil Nadu.
  - More evaporation of water in arid-semi arid region.
  - Flooding of sea water in cultivated areas.
  - Highly accumulation of Na salt causes the sodic soil.

- These soils contain high pH, poor infiltration rate and poor permeability, poor physical condition which offer high mechanical impedance to the growth of roots.

- **Effects of salt stress on different fruit crops**
  - When concentration of the salt exceeds threshold value, growth rate and size of the plant progressively decreased
  - Plant become stunted with restricted lateral shoot level.
  - Reduction in size of the leaves and fruits
  - Decrease in fresh and dry weight of different plant parts
  - Reduction on number of leaves resulted to low yield.
  - Generally, leaves become thicker than normal
  - The top growth suppressed more than the root growth
- **When concentration of sodium salt is increased**
  - They affect the plant growth both physically and nutritionally.
  - Leaf burn and necrotic symptoms were observed in mango and grapes.
  - Yield losses in fruits are generally more because of more effect of specific toxicity than osmotic effect in fruit crops
- **Physiological effects of salt stress**
  1. Osmotic effects
  2. Specific ion effects or toxicity
  3. Nutritional imbalance
  4. Effect on germination

     Fruit crops grades for their germination.

     Date palm: Tolerant (5 to 10 dS/m)

     Fig, Apricot, Grapes, Guava and Ber: Medium tolerant (3-5 dS/m)

     Citrus, Apple & Peach: Sensitive (1.5 to 3 dS/m)
  5. Growth and development
  6. Effect on metabolism
  7. Effect on photosynthesis
  8. Effect on respiration
  9. Effect on carbohydrates
  10. Effect on nucleic acid and protein
  11. Effect on plant hormones

- **Management of salt stress in fruit crops**

1. **Natural salinity resistance**

   A. Primary stress

   I. Avoidance mechanism
   1. Salt exclusion
   2. Salt extrusion
   3. Salt dilution

   II. Tolerance mechanism
   1. Salt secretion
   2. Ion balance

   B. Secondary stress

   I. Osmoregulation
   1. Salt accumulate
   2. Organic solute accumulation

   II. Nutrient deficiency

2. **Crop improvements**
   a) Breeding aspects against salt stress
   b) Genetic engineering

3. **Soil and water management**
   a) Leaching of salts through drainage system
   b) Reclamation of salts affected soil
   c) Irrigation methods
   d) High fertility levels
   e) Improving physical properties

4. **Use of rootstock**

5. **Use of growth regulators**

6. **Other methods**

8.3 Biotic stress

- Occurs as a result of damage done to an organism by other living organism (such as bacteria, viruses, parasites, beneficial and harmful insects, weeds and cultivated or native plants).

8.3.1 Herbivore

- Low levels damage – can be completely compensated by plants in terms of fitness.
- Further increments -the intensity of damage result in a decreasing ability to maintain complete tolerance

8.3.2 Tolerance

- Photosynthetic enhancement
- Growth enhancement
- Advancing the timing of bud break
- Delaying the senescence of infected tissue:(due to increased levels of cytokinins)
- Increasing nutrient uptake

8.3.3 Strategies that limit the extent of disease in an infected host

- Barriers to infection,
- Immune response
- Rapid cell death
    - in the immediate region around the wound
    - they combat the pathogen by limiting its spread

References

Chadha, K.L. and Pandey, R. M. (1986). Psidium guajava L. Handbook of flowering. 5: 43-51.

Lunagariya, R. J., Bambhaniya, U., Zankat, S. B., Patel, R., Ganvit D. M.; Parmar, V. M., Chitroda, R. L., Kanzaria, D. R. and Thumar, K. P. (2023). Causes and remedies of fruit drop major fruit crops. International Multidisciplinary journal of applied research, 10(6): 2321-7073.

Paull, R. E. and Reyes, M.E.Q. (1996). Pre-harvest weather conditions and pineapple fruit translucency. Scientia Horticulturae. 66: 59-67.

Rao, N. S.; Shivashankara, K.S. and Laxman, R.H.(2016). Abiotic Stress Physiology of Horticultural crops.

Sakuma, F., Umeya, T., Tahira, K., Katagiri, S. and Hiyama, H. (1995). Effects of high temperature and/or gibberellin treatments during early fruit development on the occurrence of watercore in Japanese pear (Pyrus pyrifolia Nakai cv. Hosui). Journal of the Japanese Society for Horticultural Science, 64(2): 243-249.

Thornton, P. and Cramer, L. (2012). Impacts of climate change on the agricultural and aquatic systems and natural resources within the CGIAR\u2019s mandate. [Contributing authors include Vladimir Smakhtin of IWMI] (No. H045997).

Turner, D. W., Fortescue, J.A. and Thomas, D. S. (2007). Environmental physiology of the bananas (Musa spp.). Brazilian Journal of Plant Physiology,19(4): 463-484.

Turner, D.W. and Lahav, E., (1983). The growth of banana plants in relation to temperature. Functional Plant Biology,10(1): 43-53.

9

Integrated and Modern Approaches in Water and Nutrient Management in Fruit Crops

Lunagariya Radhika J., V.M. Parmar and H.J. Senjaliya

Introduction

- Successful horticultural production is not possible without sufficient moisture regimes and for that, the appropriate techniques should be adopted.
- In the total cultivated land of India, 75 % of land contributed as a rainfed area or dry land area in which 60% with semi-arid and 19.3 % of area with arid zone.
- The irrigation facilities are covered only 38 % of the total cultivated land of the country.
- The moisture is always the limiting factor in arid area.
- So, conservation of soil moisture or water management is the most important task for arid fruit production.
- Out of total cultivated area of Horticultural production, 1 lakh ha is under arid or semi-arid.
- In arid area, some fruit crops with specials features are adapted for their production. So, in this area, the soil moisture conservation or the water harvesting techniques is the prime important tools for crop production with best management systems.

Moisture conservation techniques

9.1 Moisture conservation through collecting runoff or water harvesting

9.1.1 Runoff collection

a. Farm ponds or tanks
b. Well recharge or well percolation

9.1.2 *In situ* water harvesting techniques

a. Micro-catchments
b. Crescent shaped and opening catchpits
c. Terracing in hilly areas
d. Trenches
e. Contour bunds
f. Micro relief
g. Ridges

9.1.3 Use of anti-transpirants

9.1.4 Sod culture

9.1.5 Weed control

9.2 Moisture conservation through reducing leaching

9.2.1 Increasing water holding capacity by soil sub surface barrier

9.2.2 Tillage method

a. Off seasonal tillage
b. Dead furrow system
c. Topiculteure

9.3 Moisture conservation through reducing the evapo-transpiration rate

9.3.1 Wind breaks/Shelter belts

9.3.2 Mulching

a. Plastic mulch
b. Organic mulch
c. Vertical mulch
d. Dust or soil mulch

9.4 Moisture conservation through different irrigation methods

a. Drip/sprinkler irrigation
b. Pot/Jaltrpati method
c. Other methods

9.5 Moisture conservation through collecting runoff or waterharvesting

9.5.1 Runoff collection

a. Farm ponds or tanks

Farm pond or tank is made as per the slope of the field in wich the tank of pakka structure with cover is made and the rainwater is collected in ponds or in tanks and recycle it for supplementary irrigation.

b. Well recharge or well percolation

This is also good aspect to collect the rain water in well for supplementary irrigation.

9.5.2 *In situ* water harvesting techniques

a. Micro-catchments around the tree

- It is direct using of the collected water to the root zone of the tree
- The water shade area is divided into micro catchments considered optimum either for a tree or a group of trees.
- In case of flat area, the catchments could be provided on two sides of a tree row or around each tree, where as in hilly area, the tree is planted at the lowermost corner of the micro catchments

b. Crescent shaped bunding and opening of catch pits

The crescent shaped or semicircular bunds with diameter of 6-13 m are prepared and catch pits are also dug out at the same time on the upper side of the slope. The trees are planted on the centre of the crescent. The crescent bund will collect the rainwater while the catch pits conserve the collected water. New catch pits are opened as the old ones get filled up with the silt or organic matter.

c. Planting on terracing in hilly area

In hilly areas, the terraces are made to consider the slope. The broad terraces are made with the slope towards the higher elevation and trees are planted.

d. Different types of basins

On rolling topography, the basins with shape of half-moon, V-shape, Fish bone type ridges could be made across the slope and "Saucer shape" depressions could be made for *in situ* with 15 cm depth at center and 1 m diameter on a leveled land. On mild sloping land (2-4%), the shape of " V " of 15 cm depth with a minimum arm length of 1 m are made and tree is planted at near the apex of "V".

e. **Trench planting**

It is also *in-situ* water harvesting technique in which trenches varies with 0.5 to 1.0 m deep are dug across the contours, and fruit trees such as ber, aonla, annonas, custard apple *etc.* are planted. The trenches collect the rainwater along with silt and organic matter, which promote the growth of the tree also. It is best for pineapple under dry condition.

f. **Micro-relief**

In the plateau land at Hyderabad, this system of planting was found useful. Which uses both *inside* water harvesting as well the runoff collected in a ponds for irrigation at critical periods. For this, the 3 m wide & 25 cm high ridges are made which are spaced 6 m apart. The surface of ridge is covered with tank silt to increase runoff yield. Alongside the ridge, fruit trees are planted.

g. **Other methods**

1. Different types of ridges
2. Contour bunds
3. Trenches and Bunds

Micro catchments in ber

Terracing

Contour

9.6 Moisture conservation through reducing leaching

9.6.1 Increasing water harvesting capacity of soil by adding sub surface barrier

i. Fine fraction clay, silt and organic matter.

ii. Placement of slowly permeable layer of bentonite clay at the bottom of the profile or at appropriate depth to arrest deep percolation.

iii. The other soil materials like pondsilt, exfoliated vermiculate and FYM are also improves the water holding capacity of the soil.

9.6.2 Tillage

i. Off-seasonal tillage

ii. Dead-furrow system

9.7 Moisture conservation through reducing evapo-transpiration

9.7.1 Wind breaks and shelter belts

- Wind breaks is the most important device for arid lands.
- Wind break is planted on the western and northern side
- It protects the plantation from hot wind as well as reduce the soil erosion in the area of high-density wind.
- It reduces the heavy evapo-transpirational and transpirational losses of water.
- *Cordia myxa, Zizyphus mauratiana, phoenix spp., Eucalyptus and Acasia spp.* are suitable for arid; Dalbergia, *Ailanthus excelsa*, Azadirachta *etc.* for semi-arid and *Tamarix aculeata* and *Phoenix* spp. forcoastal belts.

9.7.2 Mulching

Mulching is the artificial or natural (organic or inorganic) cover on the soil surface to minimize the evapo-transpiration from the soil. Following are the types of mulches

a. Plastic mulch
b. Organic mulch
c. Vertical mulch
d. Dust or soil mulch

Mulching is a common practice that involves applying materials to the field before, during, or soon after sowing in order to support and spread over the soil surface, such as plastic material, crop residues, livestock manure, sands, rocks, and cement (Gan *et al.*, 2008).

The main goals of mulching are to limit evaporation or water erosion (Gan *et al.*, 2013), boost soil temperature, improve the soil water supply capacity (Zhou *et al.*, 2011 and El-Beltagi *et al.* 2022), and suppress weeds (Chalker-Scott, 2007).

Mulching causes improvement in crop production, fosters plant growth, and reduces water usage (Abdrabbo *et al.*, 2017 and Yu *et al.*, 2018).

9.8 Irrigation methods used for fruit crops

a. Drip irrigation
b. Sprinkler irrigation
c. Pot irrigation method
d. Jaltripati method

References

Abdrabbo, M. A. A.; Saleh, S. M.; Hashem, F. A. (2017). Eggplant production under deficit irrigation and polyethylene mulch. Egypt J. Appl. Sci.,32: 148–161.

Chalker-Scott, L. (2007). Impact of mulches on landscape plants and the environment—A review. J. Environ. Hortic., 25: 239–249.

El-Beltagi, H. S., Ullah, I., Sajid, M., Basit, A., Shehata,W. F., Shah, S. T., Alturki, S. M., Ullah, A.; Aziz, I. and Ali, F. (2022). Influence of maturity stages on postharvest physico-chemical properties of grapefruit (Citrus paradisi var. 'Shamber Tarnab') under different storage durations. Not. Bot. Horti Agrobot. Cluj-Napoca, 50: 12620.

Gan, Y., Siddique, K. H. M., Turner, N. C., Li, X. G., Niu, J. Y., Yang, C., Liu, L. and Chai, Q. (2013). Ridge-Furrow mulching systems-an innovative technique for boosting crop productivity in semiarid rain-fed environments. Adv. Agron. 118: 429–476.

Gan, Y. T., Huang, G. B., Li, L. L., Liu, J. H., Hu, Y. G. (2008). Unique conservation tillage practices in northwest China. In No-Till Farmingm Systems, World Association of Soil and Water Conservation; Goddard, T., Zoebisch, M.A., Gan, Y., Ellis, W., Watson, A., Sombatpanit, S., Eds.; Funny Publishing: Bangkok, Thailand, pp. 429–445.

Yu, Y. Y., Turner, N. C., Gong, Y. H., Li, F. M.; Fang, C., Ge, L.J. and Ye, J. S. (2018). Benefits and limitations to straw- and plastic-film mulch on maize yield and water use efficiency, a meta-analysis across hydrothermal gradients. Eur. J. Agron.,99: 138–147.

Zhou, J. B.; Wang, C. Y.; Zhang, H.; Dong, F.; Zheng, X. F.; Gale, W. and Li, S. X. (2011). Effect of water saving management practices and nitrogen fertilizer rate on crop yield and water use efficiency in a winter wheat-summer maize cropping system. Field Crops Res., 122: 157–163.

10

Nutrient Management in Fruits Crops

Khyati Savani, Lunagariya Radhika J. and D.R. Kanzaria

10.1 What is nutrient management?

- Nutrient management is an important component in orchard management for high efficiency and high fruit quality. Fruit trees are perennial woody plants which can store nutrients and then release them for growth.
- Therefore, the nutrient uptake of fruit trees includes the total nutrients of the fruits, leaves, new branches, and time increments of storage. Nutrient uptake is different under different yield levels. For example, the N uptake of apple trees was 100–120, 110– 130,120–140, and 130–150kg/ha under yield levels of 30, 45, 60, and 75t/ha, respectively.
- Basal fertilization is considered to be important for high-quality fruit production by providing nutrients for tree growth and also improving soil structure which promotes better root development. It is usually recommended and immediately applied during the dormant season up to autumn, after fruit harvest.
- Top dressing applications are usually chemical fertilizers which are applied two to four times before bud break or prior to or after blooming and at the fruit enlargement stage, depending on the requirements of each type of fruit tree.
- Using this INM method, the total nutrient demand is calculated according to target fruit yield level and NPK fertilizers are recommended based on soil fertility index (High, Moderate, and Low soil fertility) and nutrient demand with different stages. For moderate apply trees, most of N fertilizer is applied in autumn after picking to add N storage and renew tree vigor, and most of the P fertilizer is applied in spring with flowing stage to improve flower bud quality, and most of the K fertilizer is applied in summer fruiting period to improve fruit quality. Management of high-yielding fruits using suitable pruning, and irrigation, are combined in this INM.

- Since 2005, INM has been developed in different orchard and vineyard systems, such as apple, peach, jujube, grape, cherry, mandarin orange, banana, lychee, and citrus, and demonstrated for different regions of the country (total demonstration area of 7460ha). On average, compared to farming practice, INM increased fruit yields by 11.1–18.3% while reducing chemical fertilizers (N, P_2O_5, and K_2O) by 23.1–36.2%, 10.4–17.4%, and 10.4–16.6%, respectively.
- Perennial fruit trees play an important role in carbon cycle of terrestrial ecosystems and sequestering atmospheric carbon dioxide (CO2) (Lobell *et al.*, 2005; Guimarães *et al.*, 2014). According to Wu *et al.* (2012), net C sink and C storage in biomass of apple orchard ranged from 19 to 32 Tg C, respectively, and from 230 to 475 Tg C in 20 years period, amounting to 4.5% of total net C sink in the terrestrial ecosystems in China.
- In an estimate, Lakso (2010) observed that an acre of apple orchard fixed about 20 tons of CO_2 from the air each season, and provided over 15 tons of oxygen (O_2), equivalent to over 5 billion british thermal units (BTU's) of cooling power. There will be an increasing importance of nutrient efficient cultivars that are high producers as a function of adaption on low nutrient soil conditions, including root morphology, root architecture, root exudates, nutrient transport and associated key enzymes (Iyengar *et al.*, 1988; Keshava Murthy and Kotur, 1998) using labelled fertilizers.
- During the last three decades, much research has been conducted to identify and/or breed nutrient efficient plant species or genotypes/cultivars within different fruit species but the success in releasing nutrient efficient cultivars has been limited (Srivastava and Singh, 2008c; Tandon and Tiwari, 2008).

10.2 Objectives of integrated plant nutrient management

- To reduce inorganic fertilizer requirement.
- To restore the organic matter in soil and to increase nutrient use efficiency.
- To maintain quality in terms of physical, chemical and biological properties of soil.
- To maintain the nutrient balance between the supplied nutrient and nutrient removed by plant and to improve soil health and productivity on sustainable basis.

10.3 Goal of INM

- Sustainable agricultural production incorporates the idea that natural resources should be used to generate increased output and incomes, especially for low-income groups without depleting the natural resource base.
- INM"s goal is to integrate the use of all natural and man-made sources of plant nutrients, so that crop productivity increases in an efficient and environmentally benign manner, without sacrificing soil productivity of future generations.
- INM relies on a number of factors, including appropriate nutrient application and conservation and the transfer of knowledge about INM practices to farmers through extension personnel.

10.4 Principles of INM

- In the soil, replenishment of nutrients removed by the crop
- Maintenance of organic matter content in the soil
- Avoidance of weeds, pests and diseases occurrence
- Management of problematic soil and soil pollution
- Control of soil erosion by improving soil physical properties

10.5 How to better use of INM

- **Through nutrient management**
  1. Based on soil test
  2. Knowledge of crop requirements
  3. Use correct recommendations to obtain yield without pollution
  4. Value of nutrients supplied
- **Through time management**
  1. To avoid possible fertilizer wastage
  2. Use application of fertilizer at right stage of crop growth
  3. By adopting 'little and often' approach to fertilizer application
- **Through choice of balanced fertilizer**
  1. Use of balanced fertilizer to meet accurately all nutrient requirement.
- **Consult the recommendations**
  1. Use the product as per the recommendation set out for application to avoid pollution and to meet the crop requirements.

- **Through integrated nutrient management**
  1. Organic manures
  2. Chemical fertilizers
  3. Bio fertilizers

10.6 Different sources of nutrient management

1. **Organic Manure:**
   - FYM & Green manure
   - Oil cakes
   - Vermicompost
   - Bio compost
2. **Chemical Fertilizers:**
   - Nitrogenous fertilizers
   - Phosphatic fertilizers
   - Potassic fertilizers
3. **Bio-fertilizer:**
   - Rhizobium
   - Azospirillum
   - Azotobacter and PSB

10.7 Advantages of INM

- Enhances the availability of nutrients.
- Synchronizes the nutrient demand of the crop.
- Provides balanced nutrition to crops and minimizes the antagonistic effects resulting from hidden deficiencies and nutrient imbalance.
- Improves and sustains the physical, chemical and biological properties of soil.
- Improvement in the soil physical properties such as granulation, porosity, water holding capacity, drainage and aeration *etc.*
- Balanced nutrition to the crops, better nutritional uptake and hence better plant growth and yield.

10.8 Advances in fertilizers and pesticides applications

10.8.1 Aerial application

- Aerial application, or what was formerly referred to as crop dusting, involves spraying crops with crop protection products from an agricultural aircraft.
- Planting certain types of seed are also included in aerial application. The specific spreading of fertilizer is also known as aerial topdressing.
- Agricultural aircraft are highly specialized, purpose-built aircraft.
- Agricultural aircraft are often powered by turbine engines of up to 1500 hp and can carry as much as 800 gallons (3,000 l) of crop protection product.

10.8.2 Injection into soil

- Liquid fertilizers for injected into the soil by injections directly to the root zone, it may be either pressure or non-pressure types.

  Procedure: Soil Injection involves the injection of liquid solutions at regular spacing in the soil throughout the area directly beneath the canopy, starting from the trunk and extending just beyond the drip line. Injections are made under pressure of about 150 to 200 psi to a depth of 8 to 10-inches.

10.8.3 Leaf color chart method for N fertilizers application

- The LCC is an innovative cost-effective tool for need based N management.
- LCC is easy to use and simple alternative to chlorophyll meter/SPAD meter (Soil and Plant Analysis Development).
- It measures leaf color intensity that is related to leaf N status.
- It is manufactured with 4 colors called Four Panel LCC and 6 colors called Six Panel LCC.
- LCC is provided with water proof laminated instruction sticker in the regional language.

10.8.4 Trunk injection

- It is a method in which fertilizers in the form of solution are injected in to the xylem vascular tissues.
- Trunk injection is the most effective alternative to spraying or soil applied pesticides for tree insect or pest control.

10.8.5 Quadcopter UAV based fertilizer and pesticide spraying system

- A remote-controlled UAV (Unmanned Aerial Vehicle) is used to spray the pesticide as well as fertilizer.
- This model is used to spray the pesticide content to the areas that can't easily accessible by humans.

- The Universal Sprayer system is used to spray the liquid as well as solid contents which are done by the universal nozzle.
- Multispectral camera is used to capture the remote sensing images which are used to identify the green fields as well as the edges of crop area.

  Drones: Drones are used to take high quality images. Drones collect information largely based on the light reflected by the crop below.It is used in crop monitoring (water, nutrient, disease, pest) an also for chemical application.

  Robots: Self steering tractors have existed for some time now-a-days Agricultural Robots also known as Agbots.

10.8.6 Controlled release fertilizers

- Fertilizers that contain a plant nutrient in particular form, the plant cannot absorb instantly.
- Uptake is delayed after application, so that controlled release fertilizers provide the plant with nutrients for a longer time in comparison to quick release fertilizers, such as urea.

- They are typically coated or encapsulated with inorganic or organic materials that manage the rate, pattern, and duration of plant nutrient release.
- The release rate of a controlled release fertilizer is designed in a pattern synchronized to meet changing crop nutrient requirements.
- Example: Polymer-coated urea

10.8.7 Nano-fertilizers (Smart nutrient delivery system)

- Nano fertilizers are synthesized or modified form of traditional fertilizers, fertilizers bulk materials or extracted from different vegetative or reproductive parts of the plant by different chemical, physical, mechanical or biological methods with the help of nanotechnology used to improve soil fertility, productivity and quality of agricultural produces.
- Nanotechnology is nothing but the study of science, engineering, and technology at the Nano scale, which is measured in manometers. Many people believe that current technologies will meet the world's expanding food demand while also providing a slew of environmental, health, and economic benefits (Wheeler *et al.,* 2005). Nano fertilizers are one of the highly engineered inputs discovered and proved to be sound in the field of agriculture (Mejías *et al.,* 2021).
- Zinc is one of the commonly deficient micronutrients in soil (Stella *et al*., 2010). Chlorophyll formation, fertilization, pollen function, and auxin synthesis all need zinc-containing nanomaterials. Zn is one of the elements that defend plants from drought conditions (Sharma *et al*., 2009).
- ZnO NPs were shown to improve chlorophyll content and protein synthesis, rhizospheric microbial activity, acid phosphatase, alkaline phosphatase, and phytase activity in cluster bean rhizosphere by Raliya *et al.* (2017).

10.8.8 Nanoherbicides

- Nanoherbicides are formulated by exploiting the nanotechnological potential for effectual delivery of chemical or biological pesticides with the help of nanosized preparations or nanomaterials-based herbicide formulations.
- Nanomaterials or nanostructures materials-based formulations could improve the efficacy of the herbicide, enhance the solubility and reduce the toxicity in comparison with the conventional herbicides.

- The development of specific herbicide molecule encapsulated with nanoparticle aims at specific receptors present at the root of the targeted weed.
- The developed nanoparticle enters the root system of the weed and gets translocated to perform its action which in turn inhibits the glycolysis of the plant root system.
- The targeted action creates starvation of the plant and thus kills it.
- These nanoherbicides could also be used in rain-fed areas where herbicides get dissipated through vaporization due to insufficient soil moisture.
- Nanoparticles can act as good carrier and also can form nanoformulation when added with herbicides.
- The nanoparticle systems for herbicide delivery are mostly composed of polymeric substances which are biodegradable with non-toxic metabolites.

10.8.9 Nanopesticides

- The recent development of a nanoencapsulated pesticide formulation has slow releasing properties with enhanced solubility, specificity, permeability and stability.
- These assets are mainly achieved through either protecting the encapsulated active ingredients from premature degradation or increasing their pest control efficacy for a longer period.
- Formulation of nanoencapsulated pesticides led to reduce the dosage of pesticides and human beings exposure to them which is environmentally friendly for crop protection.
- So, development of non-toxic and promising pesticide delivery systems for increasing global food production while reducing the negative environmental impacts to ecosystem

References

Guimarães, V.; Danielle; Maria I. S.; Gonzaga José de O. and Melo Neto (2014). Management of soil organic matter and carbon storage in tropical fruit crops. Revista Brasileira de Engenharia Agrícola e Ambiental,18: 301-306.

Iyengar, B. R. V.; Keshva Murthy, S. V. and Shivananda, T. N. (1988). Seasonal variations in root quality in mandarin (Citrus reticulata Blanco). Journal of Nuclear Agriculture and Biology, 28: 107-10.

Keshava Murthy, S. V. and Kotur, K. C. (1998). Appropriate time and placement of single superphosphate in 'Alphonso' mango (Mangifera indica L.). Journal of Nuclear Agriculture and Biology,28(2): 107-10.

Lakso, A. L. (2010). Estimating the environmental footprint of New York apple orchards. New York Fruit Quarterly. 181: 26-28

Lobell, D. B.; Field, C. B.; Chahill, K. N. and Bonfils, C. (2005). Impacts of future climate changes on California perennial crop yields: Model projections with climate and crop uncertainties. Agricultural. Forest Meteorology, 141: 208-218.

Mejías, J. H.; Salazar, F. J.; Pérez, L.; Hube, S.; Rodriguez, M. and Alfaro, M. A. (2021). Nanofertilizers: a cutting-edge approach to increase nitrogen use efficiency in grasslands. Frontiers in Environmental Science, 9: 52.

Raliya, R.; Saharan, V.; Dimkpa, C. and Biswas, P. (2017). Nanofertilizer for precision and sustainable agriculture: current state and future perspectives. Journal of Agricultural and Food Chemistry, 66(26): 6487-6503.

Sharma, V.; Shukla, R. K.; Saxena, M.; Parmar, D.; Das, M. and Dhawan, A. (2009). DNA damaging potential of ZnO nano particles in human epidermal cells. Toxicology Letters, 185(3): 211–218.

Srivastava, A. K. (2012a). Integrated nutrient management in citrus. In Srivastava, A. K. (Ed.), Advances in Citrus Nutrition. Springler Verlag, Netherlands, pp.369-390.

Srivastava, A. K. and Singh S. (2008c). Citrus nutrition research in India: Problems and prospects. Indian Journal of Agricultural Sciences, 78: 3-16.

Stella, W. Y. W.; Priscilla, T. Y. L.; Djurisi, A. B.; Kenneth and M. Y. L. (2010). Toxicities of nano zincoxide to five marine organisms: influences of aggregate zinc size and on solubility. Analytical and Bioanalytical Chemistry, 396(2): 609–618.

Tandon, H. L. S. and Tiwari, K. N. (2008). Nutrient Management in Horticultural Crops. 3rd Edition. Fertilizer Development and Consultation Organisation. New Delhi, India pp. 1- 191.

Wheeler, S. (2005). Factors influencing agricultural professionals' attitudes towards organic agriculture and biotechnology (Doctoral dissertation, ANU, Canberra). Journal of Agricultural and Food Chemistry, 25(12): 128-134.

11

Total Quality Management in Fruit Crops

J.S. Parsana, D.K. Varu and Lunagariya Radhika J.

11.1 Quality

- Quality of fruit or any other commodity is combination of attributes, properties or characteristics like appearance, texture, flavor and nutritive value that determine their value to the consumer.
- Keeping quality is also an important quality attributes in all fruits.
- **Quality parameters**
  - Appearance (size, shape, color and freedom from defects)
  - Texture (firmness and juiciness)
  - Flavors (swetness, sourness, bitterness, astringency, aroma and off-flavour)
  - Nutritional (vitamins and minerals)
- **Quality management**

  Quality management is a method of ensuring the effectiveness of all the factors those implemented to improve and maintain the quality in acceptance form at consumer's level.

11.2 Pre-harvest management of quality

11.2.1 Genotype and rootstock management

- Genotype and Rootstocks have significant influence on fruit quality.
- Within each commodity, there is a range of genotypic variation in composition, quality and postharvest life potential.
- Plant breeders have been successful in selecting pineapple cultivars with higher contents of ascorbic acid, carotenoids and sugars.
- For example, the fruits on the dwarfing guava rootstocks *Psidium pumilum* had the higher total soluble solids and total sugars.
- Trees on *P. cujavillis* produced the fruits with the highest ascorbic acid content.

- Apple rootstocks M-779 and Crab C resulted in more sugar acid ratio in Jonathan cultivar there by improving the fruit quality.
- Coorg mandarin fruits are of better quality on Cleopatra mandarin than on Rangpur lime and *Troyer citrange*.
- Kinnow fruits have better TSS on *Troyer citrange* than on Karna Khatta and Soh Sarkar rootstocks.
- Rootstocks also affect the storage quality of the fruits. For example, the fruits produced on the dwarfing apple rootstocks have poor storage life as compared to those produced on standard rootstocks.

11.2.2 Water management

- Severe water stress results in increased sunburn of fruits, irregular ripening of pears and tough and leathery texture of peaches.
- Moderate water stress reduces fruit size and increases contents of soluble solids, acidity and ascorbic acid content.
- On the other hand, excess water supply to the plants results in cracking of fruits (such as cherries), excessive turgidity leading to increased susceptibility to physical damage, reduced firmness, delayed maturity and reduced soluble solids content.
- Therefore, application of water in proper quantity and at proper stage is helpful for improving the quality of fruits.

11.2.3 Nutrient management

- High calcium content in fruits has been related to longer post-harvest life as a result of reduced rates of respiration and ethylene production, delayed ripening, increased firmness and reduced incidence of physiological disorders and decay.
- In contrast, high nitrogen content is often associated with shorter post-harvest life due to increased susceptibility to mechanical damage, physiological disorders and decay.Increasing the nitrogen and /or phosphorus supply to citrus trees results in somewhat lower acidity and ascorbic acid content in citrus fruits, while increased potassium fertilization increases their acidity and ascorbic acid content.
- There are numerous physiological disorders associated with mineral deficiencies.For example, bitter pit of apples, cork spot in apples and pears, red blotch of lemons is associated with calcium deficiency in these fruits.

- Boron deficiency results in corking in apples, apricot and pears, lumpy rind of citrus fruits, malformation of stone fruits and cracking of apricots. Poor color of stone fruits may be related to iron and/or zinc deficiencies.
- Excess sodium and/or chloride (due to salinity) results in reduced fruit size and higher soluble solids content.Therefore, proper supply of nutrients is necessary to improve and maintain the fruit quality.

11.2.4 Pre-harvest treatments with chemicals

- Pre-harvest treatment with chemicals like plant growth regulators, fungicides and other chemicals like calcium salts and polyamines *etc.* have profound influence on post-harvest qualities including delay in ripening, reducing post-harvest physiological disorders, alleviation of chilling injury, *etc.*
- Pre harvest sprays of the fungicides, like at periodical intervals have been recommended to reduce the post-harvest spoilage due to fungal diseases in mango and oranges.
- In grapes, calcium chloride sprays are reported to improve the shelf life.
- Application of 2.5% calcium chloride @ 50 ml/sucker in the core of pineapple plant about 390 days after planting showed better ripening and improved the overall fruit quality.

11.3 Post-harvest management of quality

11.3.1 Maturity

- Maturity at harvest is the most important factor that determines storage life and final fruit quality.Immature fruits are more subject to shriveling and mechanical damage and are of inferior quality when ripe.
- Overripe fruits are likely to become soft and mealy with insipid flavor soon after harvest.
- Any fruit picked either too early or too late in its season is more susceptible to physiological disorders and has a shorter life than fruit picked at the proper maturity.
- All fruits, with a few exceptions (such as European pears, avocadoes and bananas), reach their best eating quality when allowed to ripen on the tree.
- However, some fruits are usually picked mature but unripe so that they can withstand the post-harvest handling system when shipped long distance.

- Most currently used maturity indices are based on a compromise between those indices that would ensure the best eating quality to the consumer and those that provide the needed flexibility in marketing.

11.3.2 Ripening

- Ripening is the composite of the processes that occur from the later stages of growth and development through the early stages of senescence and that results in characteristics aesthetic and/or food quality, as evidenced by changes in composition, color, texture or other sensory attributes.
- Fruits can be divided into two groups: fruits that are not capable of continuing their ripening process once removed from the plant are known as non-climacteric fruits.
- This group includes berries (such as blackberry, raspberry and strawberry), cherry, citrus (grape fruit, lemon, lime, mandarin and tangerine), grape, litchi, pineapple, pomegranate, tamarillo.
- Fruits that can be harvested mature and ripened off the plant are known as climacteric fruits.
- This group includes apple, pear, quince, persimmon, apricot, nectarine, peach, plum, kiwifruit, avocado, banana, mango, papaya, cherimoya, sapota, guava and passion fruit.
- Fruits in group two must be ripened, at least partially, before cutting to assure better flavor quality in the fresh cut products.
- Once fruits are ripened, they require more careful handling to minimize bruising.

11.3.3 Harvesting

- Method of harvesting (manual or mechanical) and time of harvesting (morning or evening), both these aspects are important.
- Generally, farmers harvest its fruits with crude methods like pulling-twisting the fruits and dropping on the ground cutting with scissors and collecting in crates/bags.
- Use of traditional scissors in harvesting leave a sharp projected pedicel persisting on fruits, tends to puncture other fruits during collection, packaging, transportation and storage.Punctured fruits get secondary infection of several pathogens like bacteria, fungi and moulds.
- This infection causes quick post-harvest deterioration in quality and quantity.

- Use of improved harvesting tools like, mango harvesters developed at the Konkan Krishi Vidyapeeth, Dapoli, Maharashtra and at the Central Institute of Subtropical Horticulture, Lucknow, who have distributed the device to the farmers in mango belts of Konkan area of Maharashtra and Uttar Pradesh and harvesting ladders to collect fruits on tree and unloading them on ground causes least damage (0.4%) to fruits.
- These precautions at harvesting minimizes post-harvest losses and help to maintain the quality upto consumers.
- Harvesting should be completed during the coolest time of the day, usually in the early morning.

11.4 Harvesting and ripening indices for important fruit crops

Fruit crop	Harvesting	Ripening
Banana	Color changes from deep green to green	Color changes from green to complete yellow
Papaya	Color changes from deep green to green	Color changes from green to yellowish green
Pineapple	150 to 160 days after flowering	Orange yellow color
Sapota	Corky brown color with intermixes yellowish tinge	Fruit become soft
Mango	Green in color, fully mature but unripe, Tapka stage	Color changes from green to yellow (TSS >16%)
Grapes	-	TSS 16%, TSS/acid ratio 30:1
Guava	Dark green to greenish yellow	-
Ber	Golden yellow color	-
Citrus	-	TSS/acid ratio 14:1 to 30:1
Pomegranate	Skin color yellowish or pink to red	-
Phalsa	Color changes to deep reddish brown and taste sweet	-

11.5 Pre-cooling

It is the process of removal of field heat from freshly harvested produce in order to reduce the post-harvest losses.

- **Methods of pre-cooling**
  - Room cooling
  - Hydro cooling
  - Forced air cooling
  - Vacuum and hydrovac cooling
  - Top icing
  - Forced air cooling is the most effective and convenient method for the pre-cooling of several fruits.
  - The pre-cooled grapes are being successfully exported to other countries under refrigerated conditions by shipment from Maharashtra and Andhra Pradesh.
  - Pre-cooling is also being practiced to some extent for the export of mangoes from Maharashtra, Andhra Pradesh and Gujarat.
- **The rate of cooling depends on number of factors**
  - Rate of heat transfer
  - Difference in temperature between the produce and cooling medium
  - Thermal properties of the produce
  - Size and shape of the produce
  - Nature of the cooling medium (Water or forced air)
  - Type of packaging
  - Stacking arrangement

11.6 Grading

- Proper grading facilities are lacking, which are required for the effective marketing of fruits through the removal of various mal adjustments prevailing in the existing agriculture marketing system.
- Grading standards are developed, by the Directorate of Marketing and Inspection, Ministry of Rural Development, Government of India, under the provision of the Agricultural Produce Grading and Marketing Act. Presently Central Agmark lab, Nagpur look after agri product grades.
- Systemic grading coupled with the scientific packaging and storage reduces the post-harvest losses and marketing costs substantially, which enables the producer to fetch a competitive price.

- It is a method of dividing products into certain groups or lots in accordance with predetermined standards. Grading follows standardization. It is a sub-function of standardization (Patil *et al.* 2014).

11.7 Quality grading

- Grading standards are also determined based on the quality characters, like color, stage of the maturity, ripening, external appearance, size, shape *etc.*
- The quality grades for different fruits are as follows.
- **Pear fruits are graded into four categories on the basisdiameter according to Agmark**

Grade	Diameter (cm)
Extra large	7.5
Large	6.5
Medium	5.5
Small	5.0

- **Apples are graded into seven grades on the basis of size (diameter)**

Grade	Diameter (cm)
Super large	8.5
Extra large	8.0
Large	7.5
Medium	7.0
Small	6.5
Extra small	6.0
Pittoo	5.5

- **Grading standards for mango, grape, pineapple and strawberry**

Fruit	Grade			
	Extra class	I class	II class	III class
Mango (weight in g)	551-800	351-550	200-350	-
Grape (bunch weight in g)	150-300	100-250	75-150	-
Pineapple (weight in g)	1500 and above	1100-1500	800-1100	<800
Strawberry (size in mm)	25 mm	22 mm	18 mm	15 mm

- **Various gardes given by Agmark in citrus fruits**

Crop	Grade	Diameter (cm)
Mosambi		
	Extra special	7.50
	Special	6.75
	Good	6.25
Mandarin		
	Extra special	8.00
	Special	7.50
	Good	6.75
	A	6.25
	B	5.25
Kagzilime		
	Extra special	4.25
	Special	4.00
	Good	3.75
	A	3.00
Grape fruit		
	Extra special	11.25
	Special	10.25
	Good	10.00
	A	9.50
	B	8.75

11.8 Technologies developed for the grading of fruits

- The National Research Center for Citrus Nagpur has developed size grader for the oranges.
- In spite of some developments, mechanical grading of the fruits in India is still in its infant stage and trading equipment has not been put into a large-scale commercial use in the field.
- **Suggestions for improving the grading practices**
- Grading standards are available for a few fruits only and there is need to frame grading standards for all the fruits.
- Proper understanding and coordination between different organizations dealing in horticultural crops.
- Studying operating models of the quality control organization and trading followed in other countries such as USA.

- Co-operatives like NDDB, Fresh Marketing Cooperative Society, Hyderabad and HOPCOMS, Bangalore could play a vital role in the grading activity of fruits at the producer's level.

11.9 Packaging

It is a fundamental tool for the post-harvest management of highly perishable commodities like fruits.

- **The important packaging requirements**
    - Protection against bruising and physical injury
    - Protection against microbial contamination and deterioration
    - Provide ventilation for respiration and exchange of gases
    - Protect against moisture/weight loss
    - Control the ethylene concentration in the package
- **Alternative packages**
    - Corrugated fibre board (CFB) boxes
    - Combination boxes
    - Jute reinforced boxes
    - Bomboo baskets
    - Corrugated polypropylene board box
    - Plastic crates
    - Wooden boxes
    - Molded pulp trays/thermoformed plastic trays
    - Stretch wrapping
    - Modified atmosphere packaging

11.10 Packaging material

1. **Corrugated fibre board (CFB) boxes**
    - CFB boxes are widely used as shipping containers for fresh produce because number of advantages over wooden containers.

- **Advantages**
    - Excellent cushioning property.
    - Low cost to strength and weight ratio.
    - Smooth and non-abrasive surface.

- Good printability on the outer surface of the board.
- Reusable and recyclable.
- Can be manufacture as tailor made design.
- Easy to set up and collapsible for storage.
- Ventilation could be provided as breathing holes in the boxes.

2. **Combination boxes**
   - They are manufactured with combination of plywood and corrugated fibre board, which have also been tried and they have a high stack load capacity.

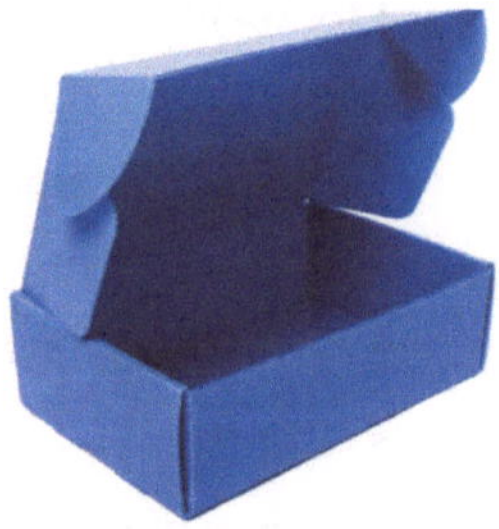

3. **Corrugated polypropylene board box**
   - It is a recent introduction in market, which is made of the corrugated polypropylene board for the purpose of multistrip package.
   - These are hygienic, sturdy, light in weight, water resistant and have a high bursting strength.
   - Corrugated boards, credited to the inventive genius of Albert L. Jones in 1871 (United States Patent Office, 1871), persistently revolutionized the packaging industry (Kouko *et al.*, 2020).
   - Lightweight, with excellent weight specific stiffness and great strength properties (Schmitt, 2019 and Chen *et al.* 2009), corrugated boards play a pivotal role in constructing robust boxes and containers for the safe transportation and protection of goods at a lower cost (Nyman, 2004).
   - The increasing concerns surrounding environmental pollution further accentuate the importance of corrugated boards. Predominantly made from cellulose (Niskanen, 2012) —a renewable resource—these packaging materials are highly ecological. They serve as a viable and preferred alternative to their non-biodegradable counterparts, mainly plastic (Adamopoulos, 2006).

4. **Plastic crates**
   - The plastic crates made of either HDPE or PP by injection moulding process have got extensive use for the packaging of fresh fruits.

- **Advantages**
  - Strong, rigid and amenable to high stack height.
  - Could be reused for number of journeys.
  - Can be cleaned easily and disinfected.
  - Available in different sizes, shapes and attractive colors.
  - Weather resistant and thus could be used in humid areas.

5. **Molded pulp trays/thermoformed plastic trays**
   - Molded pulp trays are now successfully being used for the packaging of apples.
   - Another suitable alternative to the molded pulp trays would be plastic thermoformed trays.

   - These trays with the cavities to hold an individual fruit help in holding the delicate fruit individually and avoid bruising injury.

6. **Stretch wrapping:**
   - Use of the cling plastic films for the stretch wrapping of high value fruits like pomegranate, oranges *etc.* may be considered for the retail marketing.

7. **Modified atmosphere packaging**
   - MA packaging is the process of modifying the atmospheric gaseous level (reduction of O_2 and elevation of CO_2) around the fruit inside film packaging
   - Advantages:
   - The modified concentration helps in reducing respiration rate and softening of product.

- Retards the incidence of fungal infection.
- Limit the effect of ethylene in accelerating senescence.
- Reduce the weight losses.

- **Recent developments in the modified atmosphere packaging**

I. Development of the gas permeability (MA) package films

- High ethylene vinyl acetate polyethylene films with an inherently high gas permeation rate are developed to avoid the adverse consequences of anaerobic respiration.
- Micro perforated film structures using laser perforation technologies have been successfully employed for the high respiring fruits (Brody).
- Another line of development is simple appliances made of the low permeable materials like rigid PVC, which may regulate the passage of gases through them, when fixed on the packages.

II. Development of the active packaging technologies

- Modified atmosphere (MA) conditions are created inside the packages by an active modification and also using O_2, CO_2 and ethylene scavengers within the package (Labuza).
- These scavengers may be held in small sachets within the packages or impregnated in wrappers or into porous materials like vermiculite.
- Another development in the area of active packaging is the 'smart wrap' developed in Australia, which contains a permeable membrane impregnated with an ethylene absorbent, an antifog material to avoid moisture condensation and a slow-release fungicide to inhibit the mould growth.

III. Tailoring of the plastic film laminates

- Tailoring of two or more film laminates preferably low and high barrier film would help to have gas transmission characteristics, similar to those required for the modified atmosphere packaging (MAP) of several fruits.
- The tailoring film laminates of the PVC and BOPP have increased the shelf life of apples by 2.25 times at 15°C.

11.11 Storage

- Proper storage of fresh fruits for extended periods is very much essential for the proper utilization of fruits in the season of glut.
- In tropical climate, with a high temperature and humid weather, the storability of fruits is very much reduced with heavy losses.
- Researchers have shown that by adopting the low temperature storage immediately after harvest, the metabolic activities, like rate of respiration, ethylene release, buildup of the respiration heat, thermal decomposition and microbial spoilage could be reduced with the retention of quality and freshness for a longer period.

- **Types of storage**
    - Room storage
    - Cold storage
    - Non-refrigerated air-cooled storage
    - Pit storage
    - In-situ storage
    - Controlled atmosphere storage
    - Pusa zero energy cold storage
- **CA storage**
    - The storing of product in low level of oxygen and increased level of carbon dioxide in order to slow down the rate of respiration
    - The gaseous level has to be maintained precisely throughout the storage period that should be product specific.
    - CA storage is generally accompanied with low temperature.
    - CA/MA combined with low temperature can give more beneficial results to extend the shelf life without compromising the quality.

- **Recommended CA storage requirements**

Commodity	Temperature (°C)	O_2 (%)	CO_2 (%)
Avocado	10 to 13	2 to 5	3 to 10
Banana	13 to 15	2 to 5	2 to 5
Kiwifruit	0 to 5	1 to 2	3 to 5
Citrus	10 to 13	5 to 10	0 to 5
Mango	13	3 to 5	5 to 10
Pineapple	7 to 13	2 to 5	5 to 10
Strawberry	0	5 to 10	15 to 20

- **Recommended storage conditions for important fruit crops**

Commodity	Temperature (°C)	Relative humidity (%)	Approximate shelf life
Apple	-1 to 4	90 to 95	1-12 months
Banana	13 to 15	90 to 95	1-4 weeks
Grapes	-1 to 0.5	90	1-2 months
Guava	5 to 10	85 to 90	2-3 weeks
Citrus	10 to 13	98 to 100	1-6 months
Mango	13	85 to 90	2-3 weeks
Papaya	7 to 13	85 to 90	1-3 weeks
Pineapple	7 to 13	85 to 90	2-4 weeks
Pomegranate	5	90 to 95	2-3 months
Strawberry	0	90 to 90	2-3 weeks

11.12 Pusa zero energy cold storage

- This on-farm storage chamber made of bricks was developed by Indian Agricultural Research Institute, New Delhi and it works on evaporative cooling.
- The temperature inside the chamber is 10 to 15 °C lower than the ambient and relative humidity is around 90%.It can be used to store a number of fruits on farm.
- **Storage life of fruits in Pusa Zero Energy Cool Chamber**

Crop	Cool chamber		Room temperature	
	Days	Weight loss(%)	Days	Weight loss (%)
Mango	9	5.0	6	14.9
Banana	20	2.5	14	4.8
Grape	70	10.2	27	11.9
Sapota	14	9.5	10	20.8
Lime	25	6.0	11	25.0

Kinnow mandarin	60	15.3	14	16.1
Pomegranate	48	4.5	30	11.2

References

Adamopoulos, S. (2006). Fiber analysis techniques for sustainable manufacturing of corrugated board and packaging. Manag. Dev. Mt. Isl. Areas.

Chen, L.; Xiao, S.; Li, Y.; You, S. (2009). Modeling and Simulation of Pre-compressed Multilayered Corrugated Board during Impact. In Proceedings of the 2009 International Conference on Computer Modeling and Simulation, Macau, China, pp. 143–146.

Kouko, J.; Lehto, J.; Tuovinen, T.; Vahala-Partanen, P. (2020). On the founding of the corrugated board industry and the introduction of box compression testing: 1871–1914. In Proceedings of the Progress in Paper Physics Seminar, Jyväskylä, Finland.

Niskanen, K. (Ed.) (2012). Mechanics of Paper Products; De Gruyter: Berlin, Boston.

Nyman, U. (2004). Continuum Mechanics Modelling of Corrugated Board; Lund University: Lund, Sweden.

Patil, R. T.; Nanda, S. K.; Kadam, D. M. and Narula, A. M. (2014). Entrepreneurship development: a model agro processing centre. In: Mangaraj S, Patil RT and Chandra P. (Eds.). Agro Processing and Value Addition for Entrepreneurship Development. Satish Serial Publishing House, New Delhi, India, pp. 1-12.

Schmitt, C. (2019). Investigation of the Influence of Fibre Sources on the Strength of Corrugated Boxes and Prediction of Final Strength Properties from Basic Furnish Properties. Ph.D. Thesis, Technische Universität Darmstadt, Darmstadt, Germany.

United States Patent Office. Improvement in Paper for Packing. U.S. Patent 122023A, 19 December, 1871.

12

Standards of Major Fruit Crops

V.M. Parmar, M.S. Tondihal and Subhrajyoti Mishra

12.1 Mango

Botanical name	***Mangifera indica***
Family	Anacardiaceae
Origin	S-E Asia
Ch. No.	40
Inflorescence	Panicle
Pollination	Cross pollination
Bearing habit	Terminal
Growth pattern	Single sigmoid growth
Type of fruit	Drupe
Fruit weight	330 g
Fruit length	8.9 cm
Fruit width	7.7 cm
Specific gravity	1.0 to 1.02
Weight of stone	40.5 g
Length of stone	11.19 cm
Width of stone	4.8 cm
Viability of stone	3 weeks
Skin weight	19.7gm
Pulp weight	239.27 g
Peel: Pulp ratio	6.20
Stone: Pulp ratio	3.93
TSS	17.74%
Acidity	0.23%
NRS(Non-Reducing Sugar)	14.04%
RS(Reducing sugar)	4.94%
Vit. A contain/100gm	4800 I.U.
Vit. C contain/100gm	16 mg
β- Caroteine contain/100gm	2.4mg

Self-life	35 days
Climacteric/Non-climacteric	Climacteric
Edible part	Mesocarp
Time of flowering	Nov. – Dec.
Days taken to maturity after fruit set	4 months
Yield	80kg/tree

12.2 Sapota

Botanical name	***Manilkara zapota***
Family	Sapotaceae
Origin	South Mexico
Ch. No	26
Inflorescence	Solitary
Pollination	Cross pollination
Type of fruit	Berry
Fruit weight	115gm
Fruit length	5.25 cm
Fruit width	4.3 cm
Specific gravity	1.02 – 1.05
Weight of seed	1.64gm
Length of seed	2.0cm
No. of seed in 100 gm	60
Skin weight	7.27gm
Pulp weight	73.95gm
Seed: Pulp ratio	45.55
TSS	21.40%
Acidity	3.5%
NRS(Non-Reducing Sugar)	2.2%
RS(Reducing sugar)	9.6%
Vit. A contain/100gm	978 I.U.
Vit. C contain/100gm	6.75mg
β – Caroteine contain/100gm	97 microgram
Self-life	25 days
Climacteric/Non-climacteric	Climacteric
Edible part	Mesocarp
Time of flowering	Oct.-Nov. & Feb.- Mar.
Days taken to maturity after fruit set	4-5 month
Yield	166kg/tree

12.3 Papaya

Botanical name	***Caricapapaya***
Family	Caricaceae
Origin	Tropical America
Ch. No	18
Bearing habit	Axillary
Inflorescence	Solitary
Pollination	Cross pollination
Growth pattern	Single sigmoid growth
Type of fruit	Berry
Fruit weight	1.75kg
Fruit length	32cm
Fruit width	15cm
Specific gravity	1.03
Viability of seed	45 days
No. of seed in 100 gm	4570
Skin weight	76.33gm
Pulp weight	787.5gm
TSS	7-11%
Acidity	0.91%
NRS(Non-reducing Sugar)	3.44%
RS(Reducing sugar)	5.05%
Vit. A contain/100gm	2020 I.U.
Vit. C contain/100gm	57mg
β - Caroteine contain/100gm	57.27mg
Papain content	1500-1600kg/ha
Self-life	5days
Climacteric/Non-climacteric	Climacteric
Edible part	Mesocarp
Time of flowering	Throughout the year
Days taken to maturity after fruit set	145-165days
Germination %	95.8
Number of days to take the germination	2-3 week
Age of seedling	2 months
Height of seedling	15 cm
No. of nodes	8-12
No. of leaves	8-12
Leaf area	38.58cm^2
Collar diameter	4.16mm

Root length	7.26cm
Fresh weight of seedling	4.39g
Dry weight	0.59g
Yield	48kg/plant

12.4 Guava

Botanical name	***Psidium guajava***	
Family	Myrtaceae	
Origin	Tropical America	
Ch. No.	22	
Bearing habit	Axillary	
Inflorescence	Solitary	
Type of fruit	Berry	
Fruit weight	160gm	
Fruit length	7.8cm	
Fruit width	6.5cm	
Specific gravity	1.01-1.02	
No. of seed /fruit	375	
Viability of seed	1 year	
No. of seed in 100 gm	5882	
Skin weight	3.28gm	
Pulp weight	52.6gm	
Weight of 100 seed	1.70 gm	
TSS	12-14%	
Acidity	0.37%	
NRS(Non-Reducing Sugar)	3.57%	
RS(Reducing sugar)	5.41%	
Vit. A contain/100gm	200-400 I.U.	
Vit. C contain/100gm	300mg/100g	
β – Caroteine contain/100gm	52.12	
Pectincontent	1.06%	
Self-life	20 days	
Climacteric/Non-climacteric	Climacteric	
Edible part	Thalamus and pericarp	
Time of flowering	Ambe bahar	Feb.-Mar.
	Mrig bahar	June-July
	Hasta bahar	October
Days taken to maturity after fruit set	5-6 months	
Yield	300kg/tree	

12.5 Ber

Botanical name	***Ziziphus mauritiana***
Family	Rhamnaceae
Origin	China
Ch. No.	48
Bearing habit	Axillary
Pollination	Cross pollination
Type of fruit	Drupe
Fruit weight	13.7gm
Fruit length	3.78cm
Fruit width	2.63cm
Specific gravity	0.97
Weight of stone	1.21gm
Length of stone	2.57cm
Width of stone	0.91cm
Viability of stone	12 months
No. of seed in 100 gm	83
Stone: Pulp ratio	17.65
TSS	18.2%
Acidity	0.23%
NRS(Non-Reducing Sugar)	5.6%
RS(Reducing sugar)	4.51%
Vit. A contain/100gm	81μg
Vit. C contain/100gm	114.3mg
β – Caroteine contain/100gm	4mg
Self-life	30-40 days
Climacteric/Non-climacteric	Non-climacteric
Edible part	Pericarp
Time of flowering	September
Days taken to maturity after fruit set	150-175 days
Yield	80-200kg/tree

12.6 Pomegranate

Botanical name	***Punica granatum***
Family	Punicaceae
Origin	Iran
Ch. No.	18
Bearing habit	Axillary
Pollination	Cross pollination

Type of fruit	Balausta	
Fruit weight	295gm	
Fruit length	6.42cm	
Fruit width	7.01cm	
Skin thickness	2.16mm	
Skin percentage	30.3%	
Seed percentage	14.95%	
Juice percentage	41.24%	
Aril length	10.53mm	
Aril diameter	6.72mm	
Seed length	6.88mm	
Seed diameter	2.58mm	
No. of seed in 100 gm	163	
TSS	17.5%	
Acidity	1.85%	
Vit. A contain/100gm	90mg	
Vit. C contain/100gm	43mg	
Thiamine/100gm	0.27mg	
Riboflavin/100gm	0.25mg	
Pectin contain/100gm	1.4%	
Self-life	7 months	
Climacteric/Non-climacteric	Non-climacteric	
Edible part	Juicy aril	
Time of flowering	Ambe bahar	Feb.-Mar.
	Mrig bahar	June-July
	Hasta bahar	October
Days taken to maturity after fruit set	6-7 months	
Yield	22.4kg/tree	

12.7 Custard apple

Botanical name	***Annona squamosa***
Family	Annonaceae
Origin	Tropical America
Ch. No.	14
Pollination	Cross pollination
Bearing habit	Axillary (old season growth)
Growth pattern	Double sigmoid growth
Type of fruit	Aggregate berries
Fruit weight	233.02 gm

Fruit length	9.5cm
Fruit width	3.38cm
Number of fruit /trees	104 no.
Weight of seed	5.4gm
No. of seed /100 gm	18 – 20
Length of stone	20.50mm
Width of stone	7.70mm
Pulp weight	114.46
TSS	23.36%
Acidity	0.16%
NRS(Non-Reducing Sugar)	3.78%
RS(Reducing sugar)	16.04%
Vit. A contain/100gm	16.63 μg
Vit. C contain/100gm	28.15mg
Vit. B-6	10%
Self-life	1-2 days after ripening at room temp.
Climacteric/Non-climacteric	Climacteric
Edible part	Pericarp
Time of flowering	August to October
Days taken to maturity after fruit set	104 -108 days
Yield	13.34kg/tree

12.8 Citrus

Botanical name	***Citrus spp.***
Family	Rutaceae
Origin	China
Chromosome No. (2n)	18
Climatic adaptability	Subtropical
Fruit morphology	Hesperidium
Rate of respiration	Non- climacteric
Relative salt tolerance	Highly sensitive
Pollination	Selfpollination
Aroma	Citral
Acid present	Citric acid
Harvesting period	Feb. – March (central) Dec. – April (S.I) Dec. – Feb. (N.I)
Storage temp.	8- 10°C
Inflorescence	Cymose

Pollination	Insect (Entomophilous)
Edible part	Juicy placental hairs
Ethylene production	very low (<0.1 μL C_2H_4/ kg/ hr)
Respiration rate	very low (< 5 mg release of CO_2)
Bearing Habit	Mix bearing (terminal and axillary)
Continuation of growth	Evergreen
Based on botanical relationship	Dicot

12.9 Citrus *spp.*

Crop Name	Mandarin	Acid lime	Lemon	Sweet orange
Botanical name	*Citrus reticulata*	**Sweet lime** *Citrus limetoides* **Tahiti lime** *Citrus latiifolia* **Rangpur lime** *Citrus limonica***Kagzi lime** *Citrus aurantiifolia*	*Citrus lemon*	*Citrus sinensis*
Average fruit weight	130-200 g	35-45 g	53-80 g	150-250 g
No. of fruit/tree	350-500 (300-400 -Kinnow)	400-700	350-550	250-550
Fruit retention	55-70 %	35-50%	29-51 %	30-50%
Yield	12-37 kg/ plant (30-70 -Kinnow)	17-30 kg /plant	15-27 kg/plant	54-60 kg/plant
Average fruit length	5.5-8.3 cm	2-5cm	4.9-5.5 cm	5-7.5 cm
Average fruit width	6.0-7.6 cm	2-3 cm	4.5-5.5 cm	4-7 cm
Peel thickness	4-5 mm (0.35-0.55 cm -Kinnow)	0.25-0.30 cm	0.20-0.30 cm	0.30-0.45 cm
Pulp weight	150-180 g	25-35 g	40-65 g	120-230 g
Peel percentage	12-25%	13-21 %	10-20 %	20-30 %
Peel weight	10-12 g	7.9-10 g	7.7-9.5 g	12-20 g
No. of sacs/fruit	6-12	5-10	5-12	-
Volume of fruit	140-210 cc	36-44 cc	30-50 cc	200-300 cc
Average seed weight/fruit	0.95-2 g	1-2 g	0.9-1.8 g	1-2.3 g

Average seed per fruit	4-10	9-11	4-10	6-14
Juice contains	1-60 %	46-55 %	42-48 %	45-70%
TSS	4-7 °Brix (6.6 to 8.9 °Brix -Kinnow)	9.20-10.5°Brix	7-8 °Brix	8-12 °Brix
Acidity	1.1-1.5 %	6.5-8.8%	5-6%	5-8%
TSS - acid ratio	1.00: 1.62	1.05: 1.5	1.01: 1.60	1.2: 1.65
Ascorbic acid	20-30 mg/100 ml juice	25-30 mg/100 ml juice	45-51mg/100ml of juice	40-55 mg/100ml juice
Malic acid	-	0.20 g/100 ml (juice)	-	-
Citric acid	0.07-0.11 g/100 ml (juice)	0.08-0.10 g/100 ml (juice)	0.08-0.20 g/100 ml (juice)	0.07-0.25 g/100 ml (juice)
Storage temperature	8.0-10.00 °C at 85 to 95 % RH	8.3-10.00 °C at 85 to 95 % RH	8.0-10.00 °C at 85 to 95 % RH	7.5-10.00 °C at 85 to 95 % RH
Storage life	1-8 months	6-8 months	4-8 months	5-10 months
Germination	Around 30 - 40 days after seed sowing	25 – 30 days after seed sowing	Around 30 - 40 days after seed sowing	Around 35 - 50 days after seed sowing
Germination percentage	70-80 %	70-80 %	70-80 %	50-70 %
Height of seedling at 120 days	Around 20-30 cm	Around 15-25 cm	Around 20-30 cm	Around 20-40 cm
Leaf stalk length at 120 days	7-10 mm	8-12 mm	6.5-10 mm	8-12 mm
No. of secondary roots at 120 days	14-17	15-20	12-16	13-20
Length of seedling at 120 days	-	25-40 cm	-	-
Thickness of primary roots at 120 days	1.50-2.2 mm	1.55-2.00 mm	1.45-2.2 mm	1.45-2.5 mm
Top girth of seedling at 120 days	-	0.75-0.97 cm	-	-
Middle girth of seedlingat 120 days	-	0.75-1.1 cm	-	-

Bottom girth of seedling at 120 days	0.80-1.5 cm	0.88-1.5 cm	0.9-1.7 cm	0.86-1.6 cm
Fresh weight of seedling at 120 days	16-19 g	15-20 g	14.5-20 g	15.50-23 g
Dry weight of seedling at 120 days	7.5-11.69 g	7.65-11.58 g	8-10.75 g	7.91-13.00 g
Total leaf area of seedling at 120 days	6.8-9 cm^2	6.5-8.2 cm^2	6.2-7.89 cm^2	6.7-9.21 cm^2
Survival percentage	70 %	65-75 %	67-80%	75%
No. of leaves/ seedling	10-15	5-12	8-10	10-17
Initial fruit set	60-75%	55-80%	49-85%	50-75%
No. of seed per 100 grams	150-220	130-200	150-230	170-250
Viability of seeds	4-5 weeks	1-3 weeks	2-3 weeks	3 -4 weeks
Time of flowering	Oct.-Dec.	Sept.-Nov.	Winter month	Oct.-Jan.
Transplanting	June-July	June-July	June-July	June-July,Feb.-March
Age of seedling	Around 6-10 months	Around 6-10 months	Around 6-10 months	Around 6-12months
Seedling height	Around 30-50 cm	Around 30-50 cm	Around 30-50 cm	Around 30-50 cm

12.10 Coconut, Peach and Plum

Crop name	**Coconut**	**Peach**	**Plum**
Botanical name	*Cocus nucifera*	*Prunus persia*	*Prunus domestica*
Family	Arecaceae (Palmae)	Rosaceae	Rosaceae
Origin	South east Asia	China	Japan
Chromosome No. (2n)	32	16	16 (Japanese) 48(European)
Climatic adaptability	Tropical	Temperate	Temperate
Propagation	Seed	Seed andT-Budding	Budding and Grafting
Fruit morphology	Drupe	Drupe	Drupe
Rate of respiration	-	Climacteric	Climacteric
Relative salt tolerance	High	Low	Low

Pollination	Tall var.- Cross pollinationDwarf var.- Self pollination	Self pollination	Cross pollination
Acid present	-	-	Mallic acid
Harvesting period	All round year	May – July	June – Sept.
Storage temp.	-	0 – 0.3 °C	0°C
Storage life	-	2 -4 weeks	2 -4 weeks
Inflorescence	Spadix	Solitary	Cymose (Fasicle)
Pollination	Wind (Anemophilous)	Insect (Entomophilous)	Insect (Entomophilous)
Edible part	Endosperm	Mesocarp and Epicarp	Mesocarp and Epicarp
Ethylene production	-	Medium	High(10 -100 µL C_2H_4/kg/hr)
Respiration rate	-	Medium	Medium
Bearing Habit	Axillary(Current season growth)	Axillary(old season growth)	Axillary(old season growth)
Fruit bud	Simple bud	Simple bud	Simple bud
Continuation of growth	Evergreen	Deciduous	Deciduous
Based on botanical relationship	Monocot	Dicot	Dicot
Average fruit weight	0.8-2.0 kg	70-186 gram	35-45 gram
No. of fruit/tree	100-130nut/palm/ year	170-200	200-250
Fruit retention	-	18-26%	14-22 %
Major color compound present	-	Flavonoids (orange-yellow)	-
Yield	80-130 nuts/palm/ year	30-40 kg	26-45 kg
Average fruit length	Before dehusking- 17 to 22 cm After dehusking- 9 to 12 cm	46-70 mm	4.00-4.5 cm
Average fruit width	Before dehusking-15 to 18 cmAfter dehusking- 9 to 11 cm	52-71 mm	4.00-4.5 cm
Peel thickness	-	0.01-0.4 cm	0.1-0.2 cm
Pulp weight	-	55-70 g	35-46 g
Pulp firmness	-	18-25 newtons	1-2.7 kg/cm^2 or 70-75 Durofel

Volume of Fruit	900-2000 ml	150-173 cm^3	100-145cm^3
Average seed weight/ fruit	-	2-7 g	4 g
Average seed per fruit	-	1	1
No. of seed per 100 grams	-	30-60	50-100
Pulp/pit ratio	-	-	24-25.5 gram
Dry matter content	-	-	12 -19 %
Extract content	-	-	10-18 °Brix
Length of stone	-	40-50 mm	30-35 mm
Width of stone	-	15- 20 mm	12-15 mm
Oil content of seeds	-	40-50 %	-
TSS	Copra: 18-22% Nut water: 7-9 %	6-13 %	13-14 %
Acidity	-	1.0-1.66 °Brix	0.5-1.3 °Brix
Soluble solids	-	11- 14 °Brix	-
Ascorbic acid:	-	120-400 mg/100 gram	-
Storage life	Semi- perishable (6-12 weeks)	4 -8 weeks (Perishable)	3-5 weeks (Perishable)
Germination	11-12 weeks	4 weeks	3-5 weeks
Germination percentage	60-70 %	30-40 %	40-55%
Fresh weight of seedling	Shoot-30%, Root-29-47 %	-	-
Dry weight of seedling	Shoot-24 %, Root-20-35 %	-	-
Survival percentage	60 – 80 %	50 – 65 %	50 – 75%
No. of leaves / seedling	6 – 8 leaves	15 – 30 leaves	15 – 30 leaves
Initial fruit set	40 – 60 %	25 -40 %	30 – 65 %
Stone weight	-	1.2-2.5 g	0.8-1.3 g
Maturity	-	-	95-97 %
Viability of seeds	1 year	1 year	7-10 months
Mesocarp	-	-	95-97 %
Soluble solids	-	-	18-22°Brix
Glucose	-	-	3.5-4.5 %
Fructose	-	-	2.5-3 %
Sucrose	-	-	1.5-5.5 %
No. of Flowers/branch	-	-	125-155

Harvested fruit / 100 flowers	-	-	6-16
Vit. C	-	-	13 % per 100 gm
Time of flowering	All round year	Early summer	Spring
Transplanting	June – Aug.	Spring	Spring
Age of seedling	9-12 months old seedling	6-8 months	4-6 months
Seedling height	Around 100-150 cm	50-80 cm	50-80 cm
Length of shoot	70-130 cm	30-50 cm	30-55cm
Length of root	33-55 cm	-	-
Root no.	1-29	-	-
No. of branches	-	2-7	5
No. of node	-	4-9	4-7
No.of leaves	2 -5 leaves/seedling	15-25 leaves/plant	10-25 leaves/plant
Copra contains per nut	175-346%	-	-
Annual copra yield per palm	12-20 kg	-	-
Copra yield per ha	2000-3600 kg	-	-
Percentage of oil in copra	66-75%	-	-
Size of nut girth	34-55 cm	-	-
Free fatty acid content	0.02-0.20 %	-	-
Thickness of kernel	0.8-1.3 cm	-	-
Volume of water	255-680 ml	-	-
Weight of husked nut	400-600 g	-	-
Number of spathe per year	11-14	-	-
Number of female flowers per year	355-480	-	-
Fully mature nut have	30-40 % coir	-	-
Coconut husk	Coir- 70 % Fibre- 30 %	-	-
The opening of the inflorescence	12-13months	-	-
No. of female flower/ spadix	30-45	-	-
No. of fruit set/ spadix	16-23	-	-
No. of harvested nut / spadix	15-20	-	-
Weight of tender nut	0.5-1 kg	-	-

Weight of mature nut	1-2 kg	-	-
Total protein	6-12 g/100 g fresh matter	-	-
Reducing sugar	0.4-0.50 g/100 g fresh matter	2.5-8.5 %	7-12 %
Non reducing sugar	1.5-8.6 g/100 g fresh matter	-	-
Acid tolerant	Tolerant to alkaline soil	Slightly tolerant to acid soil	-
Collar girth	7-12 %	-	-

Coconut

- **Selection of seed nut**
    - Collect 11-month-old nut during December to May
    - Store seeds under shade for a minimum period of 60 days prior to sowing
- **Selection of Seedling**
    - Select 9-12-month-old seedling based on following characteristics
    - Early germination
    - Rapid growth and seedling vigor
    - 6 to 8 leaves for 10-12 months old seedlings and at least four leaves for 9-month-old seedling

13

Production Technology of Major Fruit Crops

Y. Desai, Lunagariya Radhika J. and Mithapara Kinnari D.

13.1 Mango

Botanical name	*Mangifera indica*	
Family	Anacardiaceae	
Origin	S-E Asia	
Important varieties	Alphonso, Kesar, Rajapuri, Vanraj, Jamadar, Totapuri, Neelum, Dashehari, Langra **For rootstock:**	
	Polyembryony rootstocks	Sabre, Chandrakaran, Bappakai, Olour, Goa, Salem, Kurukan, Nileshwar dwarf, Ambalavi, Muvandan, Nekkare, Peach
	Kurukkan	Salt tolerant
	Muvandan, Olour	Vigorous
	Vellaikolamban	Dwarf root stock for alphonso and dashehari
	13/1	Tolerant to calcareous soil
	Bappakai	Best for Neelum variety
	Gomera	Salt tolerant
	ST-9	Vigorous
Plant propagation	• Vegetative: Veneer grafting, Softwood grafting and Epicotyl grafting during July and August month. • For Rootstock, the extracted kernels pre-soaked with GA3 @ 100 ppm produced earliest germination.	
Planting and orchard establishment	• High density (5m x 5m), (4.5 m x 4.5 m). • Ultra high-density mango (3 m x 2 m) and (3 m x 1.5 m).	

Flowering, pollination, and fruit set	• Spraying with 100 ppm of GA_3 with moderate pruning in mango increased the length of new flushes, panicle length and improved yield of mango. (Shaban, 2009). • Cycocel (Chloremquat, CCC) sprayed at 1000-4000 ppm on panicles was the best one in enhancing fruit set, followed by MH at 1500 ppm in mango at Navasari, Gujarat (Chaudhuri *et al.*, 2014). • Foliar application of 4 % KNO_3 during warm temperature condition resulted in bud break in about four weeks' time (Davenport, 2003). • Application of paclobutrazol @ 4.5 g *a.i.* per tree in the month of mid of July in soil.
Fruit growth and development	Application of NAA (50 ppm) significantly improved fruit retention, yield and fruit quality (Animesh and Ghosh, 2011).
Manuring and fertilization	• Application of 100g borax per plant with spraying of 0.5% boric acid during the month of September-October solution at peanut and marble size of fruits useful for optimum improvement in fruit quality. • Recommended dose 750-160-500 g/plant NPK
Intercultural/ special operations	Black plastic mulch with a thickness of 100 μ
Harvesting	Appearance of waxy coating, dots on the fruits, relative size of the fruit are some of the external appearances of the fruits, which indicate the maturity. Harvested by "Vedo".
Grading	Mango fruits can be grouped under three categories namely, Grade-I, *i.e.* fruit weighing over 320 g, Grade-II, *i.e.* 170 to 320 g and Grade-III, *i.e.* fruits under 170 g.
Packing	CFB boxes of 5 kg and 10 kg capacity for packing and shipping of Mango
Ripening	Mango fruits of cv. Kesar should be treated with ethrel @ 750 mg/l of water for 5 minutes and kept at room temperature to enhance the ripening and get maximum ripened and marketable fruits at 9th day. (JAU, 2020).
Postharvest Treatments and Storage	• Pre-harvest spray of fungicide like Dithane M-45 (0.2%), Bavistin 0.1%, Topsin-M (0.1%) and Captan (0.2%) is effective in controlling storage disease due to anthracnose and stem end rot in Mango varieties. • Pre-cooling is also being practiced to some extent for the export of mangoes. • Hot Water Treatment (HWT) is an effective post-harvest treatment for Mangoes.

13.2 Sapota

Botanical name	*Manilkara zapota*
Family	Sapotaceae

Origin	South mexico
Important varieties	• Kalipatti, Pilipatti, Cricket Ball, dwarfing scion cultivars PKM-1, PKM-3 • For rootstock Rayan or Khirni (*Manilkara hexandra*)
Plant propagation	• Soft wood grafting, wedge grafting • Treating the scion with IAA at 750 ppm increase the percentage of success.
Planting and orchard establishment	10 m X 10 m
Flowering, pollination, and fruit set	-
Fruit growth and development	-
Manuring and fertilization	50 kg FYM, 900 g N or 450 g N plus 8.00 kg castor cake, 450 g each of P_2O_5 and K_2O per adult tree along with recommended dose.
Intercultural/special operations	-
Harvesting	Fruit can be harvested by hand, using a pole with a basket, or using machines or platforms that place the picker close to the fruit.
Grading	-
Packing	-
Ripening	-
Postharvest Treatments and Storage	-
Processing	Drying the sapota slices of 0.5 cm thickness in solar dryer up to 33 per cent recovery to maintain quality in storage up to six months at room temperature.

13.3 Papaya

Botanical name	*Carica papaya*
Family	Caricaceae
Origin	Tropical America
Important varieties	• Honey Dew or Madhu Bindu, Red Lady, GJP-1 • Pusa dwarf, Pusa Nanha for HDP • Rainbow is the premier example of a genetically engineered horticultural crop that made it to market. It is resistant to Papaya ring spot virus (PRSV). Sun Up is another PRSV-resistant variety, which was developed through biotechnological interventions.
Plant propagation	Papaya is commercially propagated by seed and tissue culture plants.

Planting and orchard establishment	• 1.8 m X 1.8 m regular • 1.2 × 1.2 m close spacing HDP • 1.2 m X 1.2 m X 2.0 m paired row planting
Flowering, pollination, and fruit set	10 per cent of the male plants are kept in the orchards for good pollination in dioecious varieties.
Manuring and fertilization	• A fertilizer dose of 250 g N, 250 g P_2O_5 and 500 g K_2O / tree / year is recommended to be applied in six equal splits at bi-monthly intervals. • Foliar spray of zinc sulfate 24.0 g (Zn 0.5 %) and Borax 10.0 g (B 0.1 %) per liter of water during 2nd and 4th month after transplanting for getting higher yield.
Intercultural/special operations	Mulching with 100μ thick black polythene cover (Auxcilia *et al.*, 2017)
Harvesting/Grading/ Packing	After harvsting, papayas should be sorted out, pre-cooled, graded into different grades on the basis of size and colour and then stored or transported for marketing.
Ripening	Papaya can be stored at 10°C and 15°C for 16 days but 12°C has been recommended as an optimum storage temperature for 2 weeks. Storage below 10°C has been known to cause chilling injuries
Postharvest Treatments and Storage	Dipping of papayas in hot water (55°C for 2-3 min) helps to protect them from postharvest diseases during storage.
Processing	• This process of making four incisions 3mm deep in the untapped fruit surface at 3-4 days interval is repeated thrice or four times over a period of 12 to 16 days for latex collection. • Spraying papaya plants with ethrel @ 200 ppm or ethephon has been reported to significantly increase the flow of latex.

13.4 Guava

Botanical name	*Psidium guajava*	
Family	Myrtaceae	
Origin	Tropical America	
Important varieties	Allahabad safeda, L-49, Chittidar, Lalit, Sweta, Arka MridulaPant Prabhat, Red Fleshed dwarfing scion cultivars	
	Pusa Srijan (aneuploid tetrasomic, 2n+2)	Potential for dwarfing, resistant to wilt
	P. friedrichsthalianum	Dwarf, resistant to wilt, nematode
	P. molle	Resistant to guava wilt
	P. pumilum	Highly dwarf, high sugar content
	P. cujavilis	Produce largest but non-uniform and rough-skinned fruits

Plant propagation	• Soaking the seeds in cold water or in GA3 (1,000 ppm) for 24 hours improves seed germination. • Guava seeds scarified with 10% hydrochloric acid for seed sowing. • Dipping of semi-hard and hardwood cuttings with NAA (50 ppm) or IAA (100 ppm) encourages rooting. • Air layering with IBA-2500 ppm, wedge grafting.
Planting and orchard establishment	• Standard spacing for guava is 6x 6m, which accommodates 277 plants/ha. • 2.0 m (row to row) x 1.0 m (plant to plant) accommodating 5000 plants/ha. • 3.0 m (row to row) x 1.5 m (plant to plant) accommodating 2222 plants/ha. • 3.0 m (row to row) x 3.0 m (plant to plant) accommodating 1111plants/ha. • 6.0 m (row to row) x 3.0 m (plant to plant) accommodating 555 plants/ha.
Flowering, pollination, and fruit set	• Mrig bahar. • Pruning of terminal portion up to 20-30 cm length of current season growth of spring flush is effective to avoid the rainy season crop.
Fruit growth and development	-
Manuring and fertilization	Urea at 1% + K2SO4 at 2% + ZnSO4 at 0.5% + Borax at 0.3% at full bloom and after fruit set (one month later) to improve vegetative growth; nutritional status; yield and fruit quality of guava. 600:300:300g/ plant/year NPK.
Intercultural/special operations	• Exposing roots, De-blossoming, Training and pruning, Bending, • Pollarding or Dehorning for bahar management • Mulching with black polythene.
Harvesting	-
Grading	-
Packing	Wooden crates or bamboo baskets can be utilized in packaging.
Ripening	-
Postharvest Treatments and Storage	• Fruits treated with maleic hydrazide 1,000 ppm in addition to 6 per cent waxol treatment, extended the storage life of fruits at room temperature. • Dipping of fruits in solution of 1,000 ppm cycocel has been found effective for storage of guavas at room temperature for 14 days.
Processing	Fruit processors are advised to use 20 % pulp with 0.3 % acidity and 17 % TSS to prepare a good quality of guava nectar (RTS) stored up to 150 days.

13.5 Pomegranate

Botanical name	*Punica granatum*
Family	Punicaceae
Origin	Iran
Important varieties	Ganesh, G-137, Mridula, Dholka, Phule Arakta, Bhagwa and Solapur Lal
Plant propagation	• Air layering with 1000 ppm Indole Butyric acid in lanolin paste • Hard wood cutting with IAA 200 ppm or soaking basal part in 50 ppm solution for 18 hours improves rooting. • IBA 100 ppm (slow dip) and 2000 ppm (quick dip) planted in the month of January improved sprouting percentage, rooting percentage, number of roots per cutting, longest root, root weight, plant height and shoot girth. • Tissue culture.
Planting and orchard establishment	• 5 m × 5 m • High-density plantation (5 m × 2 m)
Flowering, pollination, and fruit set	NAA 10 mg/l as foliar spray during September when new flush initiation occurs
Fruit growth and development	Spraying Borax@ 0.1 to 0.2%, Spraying of GA3 in the month of June at 250ppm to reduce cracking
Manuring and fertilization	• 500 g Nitrogen, 250 g Phosphorus and 250 g Potash • First spray each of ZnSO4 @ 0.3%, MnSO4 @ 0.6%, Boric Acid (17% B) @ 0.6% should be given before flower and bud initiation *i.e.* 15-20 days after defoliation when growth started. • The second and third spray of ZnSo4 @ 0.3% and MnSO4 0.6% are applied at 30 to 60 days after peak flowering.
Intercultural/special operations	• Time to time pruning throughout its life cycle. • Ethrel is used (1-2 ml/l) as defoliant to remove leaves
Harvesting	-
Grading	• Super-sized: Fruits, which have good, attractive bright red colour, weight more than 750g each and without any spot on the peel, are graded as super-sized fruits. • King-sized: Fruits, which are free from spots, having an attractive red colour and weight between 500-750g, are graded as king-sized fruits. • Queen-sized: Fruits between 400 and 500g weight, having bright red colour and free from spots, are rated as queen sized fruits. • Prince-sized: Those fruits, which are fully ripe fruits, weighing between 300 and 400g with red colour, are graded as prince-sized fruits.

Packing	• Use of the cling plastic films for the stretch wrapping. • Corrugated fiber-board (CFB) boxes are used for packaging.
Ripening	-
Postharvest Treatments and Storage	-
Processing	-

13.6 Custard apple

Botanical name	*Annona squamosa*
Family	Annonaceae
Origin	Tropical America
Important varieties	Sindhan, Balanagar, GJCA 1
	Annona glabra: Pond apple (Tolerant to flood condition).
Plant propagation	Seed, Wedge grafting
Planting and orchard establishment	5 x 5 m and 4.5 × 4.5 m
Flowering, pollination, and fruit set	The problem of poor fruit set in custard apple can be overcome considerably by application of GA3 @ 50ppm.
Fruit growth and development	• Spray of 10 to 20 ppm NAA just prior to flowering period reduces the flower and fruit drop. • During fruit development, foliar spray of 10 ppm GA3 + 5 ppm CPPU, improves the fruit size and luster of the fruits.
Manuring and fertilization	250 g Nitrogen, 125 g Phosphorus and 125 g Potash.
Intercultural/special operations	-
Harvesting	-
Grading	-
Packing	-
Ripening	-
Postharvest Treatments and Storage	-
Processing	-

13.7 Citrus

Botanical name	*Citrus spp.*
Family	Rutaceae
Origin	China
Important varieties	
lemon	Malta, Nepali Oblong, Nepali Round, Lisbon, Willafranka, Italian, Eureka, Seedless, Seville Mayor, Pant Lemon 1

Sweet Orange	Sathugudi, Batavian, Mosambi, Malta blood red, Valencia. Washington Navel, Frost, Washington, Navelina
Acid lime	PKM 1, VRM 1 Tahithi lime, Balaji (Tenali), Vikram, Rasraj, NRCC -7, NRCC -8, Phule Sharbathi, Kagzi lime
Mandarin	Coorg mandarin, kinnow mandarin, Satsuma, Nagpur santra and Kodai Orange
Plant propagation	• Apomictic seedlings, Budding, air layering • Micro-budding is a new propagation technique • In vitro grafting technique: Micrografting • Flying dragon: Ultra dwarf root stock

Root Stock	Horticultural performance			Reaction to						
	Yield	Quality	Vigour	Root rot	Nematode	Tristeza	Exocortis	Salt	Drought	Root
Rangpur lime	G	M	G	MT	S	R	S	R	R	D
Marmalade orange	G	M	G	MT	MT	R	S	R	R	D
Rough Lemon	G	L	G	S	S	R	R	T	T	D
Cleopatra Mandarin	M	M	M	T	S	R	R	MT	S	M
Sour orange	G	G	M	R	T	HS	T	T	MT	D
Sweet lime	M	M	G	S	MT	S	S	S	S	M
Trifoliate citrange	L	M	L	R	R	R	H	HS	HS	SH

G = good, M = Moderate, L = Poor or low, R = Resistant, T = Tolerant, MT = Moderately tolerant, S = Susceptible, HS = Highly susceptible, D = Deep, M = Medium, SH = Shallow

Planting and orchard establishment	Traditional 6 x 6 m
Flowering, pollination, and fruit	• To increase fruit set, spray 2,4-D @ 20 ppm (200 mg / 10 l) during flowering.
set	• Application of paclobutrazol (PP 333) @ 2.5 to 10 g per tree inducedflowering and improved fruit set. • spraying cycocel (CCC) @ 500 ppm during September and second spray after 15 days.
Fruit growth and development	For fruit retention, spray 2, 4 – D @ 20 ppm or NAA 30 ppm (300 mg / l) after fruit set (Marble size).
Manuring and fertilization	50kg FYM 500gN 375gP 500gK
Intercultural/special operations	Spraying Zinc sulphate, Manganese sulphate, Ferrous sulphate, Magnesium sulphate each at 0.125 %
Harvesting	-
Grading	-
Packing	-
Ripening	-
Postharvest Treatments and Storage	-
Processing	-

Cultivars	**Spacing**	**References**
Acid lime cv. Pramalini	2.5 x 2.5 m	Ladaniya et al., 2019
	5.0 x 2.5 m	
Tahiti lime	4 x 1 m	Stuchiet al., 2003
Kinnow	6 x 3 m	Dalal et al., 2013
Kinnow (on Troyer citrange rootstock)	1.75 x 1.75 m	Goswami et al., 1993
Kinnow (on karna khatta rootstock)	2.25 x 2.25 m	Goswami et al., 1993
Mandarins	1.8 x 1.8 m	
Oranges	4 x 2 m	
Lime and lemons	4 x 1 m	
Grape fruit and pummelo	4 x 2 m	

Crop	**Grade**	**Diameter (cm)**
Mosambi		
	Extra special	7.50
	Special	6.75
	Good	6.25
Mandarin		

	Extra special	8.00
	Special	7.50
	Good	6.75
	A	6.25
	B	5.25
Kagzilime		
	Extra special	4.25
	Special	4.00
	Good	3.75
	A	3.00
Grape fruit		
	Extra special	11.25
	Special	10.25
	Good	10.00
	A	9.50
	B	8.75

13.8 Ber

Botanical name	*Ziziphus mauritiana*
Family	Rhamnaceae
Origin	China
Important varieties	Goma Kirti, Thar Sevika, Thar Bhubhraj, Kaithali, Umran, Gola and Banarasi Rootstock: *Z. nummularia*: Dwarf, salt tolerant. *Z. rotundifolia*: Tolerant to drought
Plant propagation	T' budding and patch budding - 500 ppm GA for seed germination
Planting and orchard establishment	7 x 7 m
Flowering, pollination, and fruit set	Spray of growth regulator such as GA3 and NAA at 10-15 ppm is reported to enhance fruit setting
Fruit growth and development	Exogenous application of growth regulators such as 2, 4, 5-T; 2, 4-D and NAA at 10-30 ppm are reported to reduce fruit drop to some extent.
Manuring and fertilization	Spray 2% KNO_3 thrice at monthly intervals in January, February and March.
Intercultural/special operations	Heading back during may.

Harvesting	-
Grading	-
Packing	-
Ripening	-
Postharvest Treatments and Storage	-
Processing	-

Index